00-PUM-2T

REDUCTION OF PATHOGENS, INDICATOR BACTERIA, AND ALTERNATIVE INDICATORS BY WASTEWATER TREATMENT AND RECLAMATION PROCESSES

by:

Joan B. Rose, Ph.D.
Homer Nowlin, Chair in Water Research
Department of Fisheries and Wildlife and Crop and Soil Sciences, Michigan State University

Samuel R. Farrah, Ph.D.
Department of Microbiology and Cell Science, University of Florida

Valerie J. Harwood, Ph.D.
Department of Biology, University of South Florida

Audrey D. Levine, Ph.D., P.E.
Department of Civil and Environmental Engineering, University of South Florida

Jerzy Lukasik, Ph.D.
Biological Consulting Services of N. Florida, Inc.

Pepe Menendez, P.E.
Florida Department of Health

Troy M. Scott, Ph.D.
Biological Consulting Services of N. Florida, Inc.

2004

WERF

The Water Environment Research Foundation, a not-for-profit organization, funds and manages water quality research for its subscribers through a diverse public-private partnership between municipal utilities, corporations, academia, industry, and the federal government. WERF subscribers include municipal and regional water and wastewater utilities, industrial corporations, environmental engineering firms, and others that share a commitment to cost-effective water quality solutions. WERF is dedicated to advancing science and technology addressing water quality issues as they impact water resources, the atmosphere, the lands, and quality of life.

For more information, contact:
Water Environment Research Foundation
635 Slaters Lane, Suite 300
Alexandria, VA 22314-1177
Tel: (703) 684-2470
Fax: (703) 299-0742
www.werf.org
werf@werf.org

This report was co-published by the following organizations. For non-subscriber sales information, contact:

IWA Publishing
Alliance House, 12 Caxton Street
London SW1H 0QS, United Kingdom
Tel: +44 (0) 20 7654 5500
Fax: +44 (0) 20 7654 5555
www.iwapublishing.com
publications@iwap.co.uk

Library of Congress Catalog Card Number: 2004109531

IWAP ISBN: 1-84339-730-7

Department of Fisheries and Wildlife; Department of Crop and Soil Sciences, Michigan State University
Department of Biology; Department of Civil and Environmental Engineering, University of South Florida.
Department of Microbiology and Cell Science, University of Florida.
Biological Consulting Services of N. Florida, Inc.

The research on which this report is based was funded, in part, by the United States Environmental Protection Agency through Cooperative Agreement No. CR-827345 with the Water Environment Research Foundation (WERF). Unless an EPA logo appears on the cover, this report is a publication of WERF, not EPA. Funds awarded under the Cooperative Agreement cited above were not used for editorial services, reproduction, printing, or distribution.

ACKNOWLEDGMENTS

The cooperation and assistance by the utilities that participated in the project is highly appreciated. The utilities that provided access to their facilities are:

City of Eustis Wastewater Department, Florida
City of Phoenix Water Services Department, Arizona
City of Santa Barbara, California
City of St. Petersburg, Florida
Hillsborough County Water Department, Florida

The following utilities contributed funds to this project through the WERF Targeted Collaborative Research (TCR) Program. These additional funds enabled the sampling regimen to be expanded and therefore added significant value to this research project. The TCR participants were:

City of Orlando, Florida
City of Phoenix Water Services Department, Arizona
City of Santa Barbara, California
City of St. Petersburg, Florida
St. Johns River Water Management District

The researchers thank the following individuals for assisting with sample collection and conduct of microbiological testing: Vasanta Chivulaka, Molly McLaughlin, Stefica Depovic, and Angela Gennaccaro. Vasanta Chivulaka assisted with statistical analyses. Amy Kamarainen assisted with report production.

Report Preparation

Principal Investigator:
Joan B. Rose, Ph.D.
Homer Nowlin, Chair in Water Research
Dept. of Fisheries; Dept. of Wildlife and Crop and Soil Sciences, Michigan State University

Project Team:
Samuel R. Farrah, Ph.D.
Department of Microbiology and Cell Science, University of Florida

Valerie J. Harwood, Ph.D.
Department of Biology, University of South Florida

Audrey D. Levine, Ph.D., P.E.
Department of Civil and Environmental Engineering, University of South Florida

Pepe Menendez, P.E.
Florida Department of Health

Jerzy Lukasik, Ph.D.
Troy M. Scott, Ph.D.
Biological Consulting Services of N. Florida, Inc.

Project Subcommittee

James Crook, Ph.D., P.E., Chair
Consultant

Charles N. Haas, Ph.D.
Drexel University

Adam Olivieri, Dr. P.H., P.E.
EOA, Inc.

Stephen A. Schaub, Ph.D.
U.S. Environmental Protection Agency

William A. Yanko
Los Angeles County Sanitation District (retired)

David W. York, Ph.D., P.E.
Florida Department of Environmental Protection

Walter Jakubowski, Ph.D.
WaltJay Consulting

Targeted Collaborative Research Participants

Patricia Anderson, Assistant Director
Kim Streeter, Senior Assistant City Attorney
City of St. Petersburg, Florida

Don Brandes, Ph.D., Senior Project Manager
Elizabeth Thomas, Water Reuse Director
St. Johns River Water Management District, Palatka, Florida

Aimée Conroy, P.E., Water Services Superintendent
Phoenix Water Services Department

Rebecca Bjork, Environmental Control Supervisor
City of Santa Barbara, California

David Sloan, Environmental Services Director
Tom Lothrop, P.E., Deputy Director of Public Works
City of Orlando, Florida

Water Environment Research Foundation Staff

Director of Research: Daniel M. Woltering, Ph.D.
Project Manager: Bonnie Bailey

ABSTRACT AND BENEFITS

Abstract:

The increasing implementation of wastewater reclamation systems necessitates that adequate controls are in place to protect the users of reclaimed water from exposure to pathogens. Regulatory compliance is based on routine monitoring of indicator organisms coupled with process performance requirements. This project was conducted to compare the effectiveness of full-scale biological treatment, filtration, and disinfection for removal of bacterial and viral indicators, enteric viruses, and protozoan pathogens. Six full-scale treatment facilities were each sampled a minimum of four times over a one-year period. The relative impacts of loading conditions, process design, and operating parameters on the removal/inactivation of a suite of nine microbial species (bacteria, coliphages, enteric viruses, and protozoan pathogens) were evaluated. Operation of biological treatment with higher levels of MLSS and longer MCRTs tended to result in increased removal of microbial indicators and pathogens. Increased virus removal was associated with biological nutrient removal and nitrification processes and parasite removal was associated with enhanced nitrification. Prechlorinated shallow sand (effective size: 0.6 mm) filters were more effective than deep bed dual media (anthracite and >1 mm sand) or monomedia (anthracite or sand) filters for reductions of bacterial indicators and viruses most predominantly due to inactivation by the disinfectant, and not physical retention by the filters. Deep-bed filtration provided slightly better protozoan removal. Hydraulic loading and filter depths impacted removal of some of the indicators. Chlorine disinfection was more effective in cases where ammonia levels were low (biological nutrient removal or nitrification facilities). Infectivity assays for *Cryptosporidium* and dye assays for *Giardia* viability suggested that the proportion of infective/viable (oo)cysts remains unchanged throughout treatment, however the concentrations of infective (oo)cysts did decrease with increasing degree of treatment. The effluent quality for bacteria, coliphage, viruses and *Giardia* cysts from the different facilities was shown to be statistically significantly different.

Benefits:

- Provides evidence that reclaimed water may contain enteric viruses and/or protozoan pathogens even after filtration and disinfection and in the absence of total or fecal indicator bacteria.
- Provides a direct comparison of the concentrations of indicator and viral and protozoan pathogens in reclaimed water exposed to different types of treatment.
- Helps to identify the relative effectiveness of biological treatment systems, filtration, and disinfection for removal of viral and protozoan pathogens..
- Results should help guide facilities toward developing improved monitoring tools to assess microbiological water quality and help to catalyze discussions on a regulatory framework focused on reclaimed water quality for viruses, parasites as well as bacteria.

Keywords: Reclaimed water, protozoan pathogens, enteric viruses, indicator organisms, biological treatment, wastewater filtration, disinfection

TABLE OF CONTENTS

LIST OF TABLES

LIST OF FIGURES

LIST OF ACRONYMS

ATCC	American Type Culture Collection
APHA	American Public Health Association
BGM	Buffalo Green Monkey
CFU	Colony forming units
CPE	Cytopathic effects
DAPI	4',6-diamidino-2-phenylindole
DIC	Differential Interference Contrast
DI	Deionized
DNA	Deoxyribonucleic acid
EPA	Environmental Protection Agency
FITC	Fluorescein isothiocyanate
FDM	Focus Detection Method
F+	Fertility fimbriae
ICC	Integrated Cell Culture
ICR	Information Collection Rule
IF	Immuno-fluorescence
IMS	Immunomagnetic Separation
ICR	Information Collection Rule
MCRT	Mean Cell Residence Time
MLSS	Mixed Liquor Suspended Solids
MS	Male-specific
MPN	Most-Probable Number
MI	Michigan
PBS	Phosphate buffered saline
PCR	Polymerase Chain Reaction
PFU	Plaque forming units
PI	Propidium iodide
RNA	Ribonucleic Acid
RT-PCR	Reverse-transcriptase Polymerase Chain Reaction
RD	Rhadbosarcoma
TC	Total coliform
U.S.	United States
U.S. EPA	United States Environmental Protection Agency
UV	Ultraviolet
WERF	Water Environment Research Foundation

EXECUTIVE SUMMARY

It is important to assess the effectiveness of reclaimed water treatment practices for control of pathogens in the final treated water with the increasing interest in the use of reclaimed water for various purposes such as irrigation, or for cooling water (eg. in urban environments), and other applications, where the public is increasingly exposed to the water. Typically, filtration and disinfection are used downstream of secondary biological treatment with a goal of producing a water that yields minimal risk of exposure to pathogens. The microbiological safety of reclaimed water is routinely assessed by monitoring indicator bacteria (fecal or total coliform) in grab samples of reclaimed water. In addition to monitoring of indicator organisms, turbidity and suspended solids are also used as indirect measures of the potential presence of microorganisms. While many facility designs are capable of meeting numerical limits for turbidity, suspended solids, and coliform bacteria, limited information is available on the effectiveness of full-scale systems for removal or inactivation of pathogens. This project was conducted to evaluate the degree to which microbial indicators and pathogens are removed/inactivated through biological treatment, filtration, and disinfection at full-scale wastewater treatment and reclamation facilities in Arizona, California, and Florida.

Specific microbial measurements conducted at strategic intervals during reclaimed water production included the following bacterial indicators: total and fecal coliforms, enterococci, and *Clostridium perfringens.* Two different methods were used to measure coliphage as virus indicators. Pathogens tested included the protozoan intestinal parasites *Cryptosporidium* and *Giardia,* and cultivatable enteric viruses. All measurements were taken at each facility in the influent, secondary effluent, filtered effluent and final disinfected effluent (finished reclaimed water).

The results and conclusions can be summarized as follows:

- The concentrations of the traditional bacterial indicators (total and fecal coliforms), alternative indicators (Enterococci, coliphage) and pathogens were fairly consistent in all influents of the six treatment facilities and were not significantly different among the facilities.
- Many wastewater treatment facilities are permitted to produce effluent that contains <1 **fecal coliform** bacteria/ 100 ml. Facilities A, B, C and F met this level in100% of disinfected effluent samples, while facilities D and E did so for only 66% and 33% of the samples. Only facility C would have met the level of <1 **total coliform** bacteria /100ml in 75% of the samples collected. The other facilities had higher levels in 40 to 80% of the samples collected.
- Cultivatable enteric viruses and protozoan pathogens (*Giardia, Cryptosporidium*, total cysts and oocysts) were detected in untreated wastewater from all six facilities. Cultivatable enteric viruses were detected in 31% of the final effluent samples and were not detected in any of the samples from facilities B and E. Protozoan pathogens were detected in some of the effluent samples from all six facilities: *Giardia* cysts were detected in 79% of the final disinfected effluent samples while *Cryptosporidium* oocysts were detected in 39% of the final reclaimed effluents. Infectious *Cryptosporidium*

oocysts were detected in 30% of the final reclaimed effluents and in four of the six facilities.

- About 10-30% of the *Cryptosporidium* oocysts remained infectious as determined via cell culture through biological treatment, filtration, and disinfection. *Giardia* cysts also remained viable (1-3% of the total IFA count) according to PI exclusion and DAPI inclusion. Infectious oocysts were not detected in any of the UV disinfected effluent samples,(detection limits of 2/100 L).
- The production of reclaimed water using secondary activated sludge processes, filtration and disinfection is not universally effective for removal of pathogens. Viruses ranged in concentration from 0.3 to 3.3 MPN PFU/100L, *Giardia* cysts ranged from 6 to 390/100 L and *Cryptosporidium* ranged from 4.6 to 114 oocysts/100 L in reclaimed water.
- Inefficiencies in secondary treatment and sedimentation result in a higher microbiological loading on filtration and may impact the effectiveness of filtration and disinfection for reduction bacteria, viruses, and protozoa.
- If coliform bacterial concentrations are used to assess microbiological water quality, facilities B and C outperformed the other facilities, while facilities A and F had intermediate performance, with D and E ranked as having the highest concentrations of these indicators. However, if detection of coliform bacteria is coupled with monitoring of enterococci, *Clostridium* and coliphage, facilities F and E outperformed the other facilities, while A and B ranked in the middle and facilities C and D had the highest frequency of detection of indicators at the detection limits used in this study.
- Based on the viruses, facilities B, C, and E ranked as having the best water quality, D and F ranked in the middle and facility A had the poorest quality. Based on *Giardia* cysts, facility F had the best water quality, facility F along with E and B also ranked the best based on *Cryptosporidium* oocysts. Facilities B, D, and E compared to A and D ranked in the middle for cysts and oocysts and facilities A and C had the poorest water quality for cysts and facility C based on oocysts.
- The use of traditional indicators as monitoring tools would not have identified the potential presence of pathogens in Facilities A and C.

Operations and Design

The data obtained in this study represented only six samples from each facility at each site after the treatment processes. Thus while this study represents the most comprehensive one of its type carried out to date, much more data are necessary to confirm the trends and results that have been observed here. The effluent quality for bacteria, coliphage, viruses and *Giardia* cysts from the different facilities was shown to be statistically significantly different, as the concentrations in the influent were not statistically different, this difference can be related to removal by the processes.

- Operation of biological treatment with higher levels of MLSS and longer MCRTs tended to result in increased removal of microbial indicators and pathogens.

- Prechlorinated shallow sand (effective size: 0.6 mm) filters were more effective than deep bed dual media (anthracite and >1 mm sand) or monomedia (anthracite or sand) filters for removal of bacterial indicators and viruses.
- Hydraulic loading and filter depths impacted removal of some of the indicators, with improved removal associated with lower hydraulic loading rates (0.6-2gpm/ft^2) and deeper filters (1-1.8m).
- UV inactivation (facility E) of bacteria and coliphage indicators removed 92 to99% of these organisms but was not as effective as the extended contact times with chlorine which achieved 99.9-99.99% inactivation (facilities C and B).
- Overall, facility C with shortest MCRTs, anthracite monomedia filtration, and lowest range of chlorine residuals had the poorest finished water quality with respect to *Cryptosporidium*, coliphage, enterococci and C*lostridium*. This was despite the fact that the facility had the longest chlorine contact times.
- Facility A had the poorest finished water quality in regard to viruses and *Giardia* (poor removal of cysts). This facility had a cloth filter, highest loading rate and lowest range of chlorine contact times.
- Facilities D and E had the worst finished water quality in regard to total and fecal coliform bacteria. The UV disinfection at Facility E was ineffective at achieving the coliform standards of nondetects for the bacterial indicators although it appeared to be effective against the viruses and should have been effective based on design against the parasites. Facility D had poorer removal by the filtration but similar disinfection contact time as facility A (which was the lowest of all facilities) thus producing a final water quality with greater concentrations of bacteria.
- Facility F produced the best overall microbial quality with the longest retention times, deepest filters (dual media), least amount of ammonia impacting disinfection (70-90 min. contact times with 4-6mg/L of residual chlorine) for parasites, viruses, and indicators. Despite this, viruses and infectious *Cryptosporidium* were still detected.
- The longer chlorine contact times in Facility B (due to prechlorination of the filter) and Facility C likely contributed to reductions of the viruses in those facilities and may be needed to achieve nondetectable levels of cultivatable viruses in the reclaimed water effluent (using current methods for virus detection).
- A high percentage of reclaimed water samples were positive for IFA total counts of cysts and oocysts, and some were positive for infectious *Cryptosporidium.* A rigorous risk assessment should be conducted to determine whether other barriers to reduce viable protozoa are needed.

Reclaimed water facilities can be designed and operated to produce high water quality. Greater attention needs to be focused on improving the efficiency of secondary treatment for removal of pathogens. The effectiveness of filtration for removal and inactivation of pathogens can be improved by strategic use of coagulation and prechlorination. Finally, disinfection efficiency is impacted by the effectiveness of upstream processes for removal of particles and other potential interferences. More robust disinfection systems are needed to compensate for upstream process upsets and inconsistencies.

Based on this study, it is evident that increased monitoring for alternative indicators and pathogens is warranted to build a better data base on the occurrence and reductions of enteric bacteria, viruses and parasites in reclaimed water. Finally the authors of this report believe that more discussion should be undertaken regarding the efficacy of process requirements versus water quality goals for reclaimed water at the national level. Reclaimed water as monitored in this study in Arizona, California, and Florida is not pathogen free and exposure of the public to these waters carries some risk, albeit this level may be very low and quite acceptable to most populations. Integration of microbiological monitoring with control factors associated with process design and operations may lead to a more robust approach for assuring the safety of reclaimed water.

CHAPTER 1.0

INTRODUCTION AND BACKGROUND

Over the past few decades, wastewater reclamation and reuse has become an integral component of water resource management in many states in the U.S. and throughout the world (Rose, 1986, Rose et al.,1996; U.S. EPA, 1992, Levine and Asano, 2004). With increasing interest in the use of reclaimed water for irrigation, cooling water, and other nonpotable applications, it is important to assess the effectiveness of current treatment practices for control of pathogens. Wastewater reclamation and reuse facilities are designed to produce reclaimed water that yields minimal risk of exposure to pathogens when coupled with specific controls on the distribution and use of the reclaimed water. Typically, the potential presence of pathogens in reclaimed water is assessed using indirect measures such as turbidity or suspended solids coupled with regular sampling for indicator organisms, such as coliform bacteria (total or fecal).

While coliform bacteria have a long history of use as indicators of microbiological safety, limited data are available on how removal of coliforms compares to removal of protozoan pathogens, viruses, or other bacteria. In addition to routine monitoring of coliform bacteria, several alternative monitoring approaches have been suggested as indicators of the presence of potential pathogens. Examples of alternative indicators include Enterococci, *Clostridium perfringens*, and F-specific coliphages. Quantitative data on microorganisms (viruses, protozoa and bacteria) in untreated wastewater and throughout various phases of the treatment process are available (Rose et al., 1996, 1999, 2001). But the degree to which data in the literature can be extrapolated to full-scale treatment systems is unknown. Differences in wastewater characteristics and treatment process effectiveness may impact the fate of pathogens through wastewater treatment facilities.

Currently, within the U.S., there are no national standards targeted at the presence of pathogenic microorganisms in reclaimed water. Originally states such as California had a goal for production of reclaimed water that was "essentially pathogen (virus) free" reduced by 99.999% and this was based on the Pomona Virus Study and ability to inactivation viruses through extensive disinfection (County Sanitation Districts of Los Angeles County, 1977; State of California, 1978). Only in the State of Arizona, had there been actually water quality standards for viruses and *Giardia* cysts (Arizona, 1984). In Florida, recent rules have recommended monitoring in attempt to build a data base on protozoa (York and Walker-Coleman, L. 2000). However, in most cases data are not available on a routine basis for enteric protozoan parasites and, in most cases, the potential presence and quantification of viruses in reclaimed water effluents.

Typically, the treatment of municipal wastewater for reclamation includes biological treatment followed by filtration and disinfection. It is likely that differences in treatment operations, varying filter designs, and disinfection approaches can produce effluents of varying quality. This study was conducted to compare pathogen reduction in full-scale treatment facilities that produce reclaimed water.

1.1 Objectives

This project was conducted to evaluate the degree to which microbial indicators and pathogens are removed/inactivated through biological treatment, filtration, and disinfection at six full-scale wastewater treatment and reclamation facilities in the United States. The five objectives of this project are:

- Compare the concentrations of bacterial indicators (total coliform, fecal coliform, enterococci, *Clostridium* perfringens), viral indicators (coliphages), human viruses, and protozoan pathogens (*Giardia, Cryptosporidium*) in untreated wastewater.
- Identify process variables that influence removal of indicators and pathogens through biological treatment systems.
- Evaluate differences in indicator and pathogen removal by filtration.
- Assess the effectiveness of chlorine (combined or free) disinfection and UV for inactivation of indicators and enteric viruses.
- Examine the viability of *Cryptosporidium* oocysts through biological treatment, filtration, and disinfection.

1.2 Microbial Indicators and Pathogens

The concept of using indicators as monitoring tools is based on providing a relatively rapid means for assessing the microbial characteristics of a sample. The principle behind the use of indicator organisms is that, while it is impractical and currently nearly impossible to test water for all possible pathogens that could be present, an indicator organism that is always found in fecal material could serve as a surrogate for detection of pathogens. Ideally, indicators should have similar environmental requirements and exhibit similar patterns of growth, die-off, and susceptibility to disinfectants and toxics as pathogens. A summary of the ideal characteristics of indicator organisms is given in Table 1-1.

Table 1-1. Ideal Characteristics of Indicator Organisms.[1]

Property	Indicator characteristics
Pathogenicity	Not pathogenic
Occurrence patterns	Present when human pathogens are present; absent when they are not
Survival characteristics	Survival characteristics should be similar to pathogen survival
Reproduction	Doesn't multiply in natural waters
Inactivation	Inactivated by treatment systems at about the same level as pathogen inactivation
Source	Only source in natural waters is fecal contamination
Cost	Assay is relatively cheap, easy to perform, and has a rapid turn-around time

[1] Adapted from Harwood, 2001

Coliform bacteria are widely used as microbial indicators for wastewater reclamation and reuse applications. In this project, total and fecal coliform concentrations were compared to other bacterial and viral indicators, enteric viruses, and protozoan pathogens. The characteristics of the microbial parameters tested in this project are compared in Table 1-2.

1.2.1 Bacterial Indicators

Bacterial indicators include total and fecal coliform, enterococci, and *Clostridium perfringens*. Because of their long history of use, coliform bacteria provide a reference point to which to compare alternative indicators. As shown in Table 1-2, the coliform bacteria are rod-shaped, Gram-negative, heterotrophic bacteria that range in length from 0.5 to 2 µm. Coliforms are facultative anaerobes, capable of aerobic respiration, anaerobic respiration, and fermentation pathways for ATP synthesis. Total coliforms consist of lactose-fermenting bacteria that do not form spores and that grow at 37°C. The fecal coliform group is a subgroup of total coliforms that is defined by its ability to grow at 44.5°C. Although *Escherichia coli* is the dominant fecal coliform in the gastrointestinal tract of mammals, members of other genera such as *Citrobacter* and *Klebsiella* can meet the operational definition of fecal coliform(LeClerc et al., 2001). The enterococci are spherical, Gram-positive aerotolerant bacteria that do not utilize oxygen for ATP synthesis. Enterococci are a subgroup of fecal streptococci and tend to be more persistent than fecal coliforms, particularly through wastewater treatment processes. *Clostridium perfringens* are rod-shaped, obligately anaerobic Gram-positive opportunistic pathogens that tend to survive longer in the environment than other bacteria due to the formation of endospores. It has been suggested that *Clostridium perfringens* could be used as a surrogate indicator for protozoan pathogens because of their spore-forming capacity and resistance to disinfection.

1.2.2 Viruses and Viral Indicators

It is estimated that over 100 human enteric viruses can be transmitted by human feces (Mara and Horan, 2003). Most of these viruses infect the gastrointestinal tract and are transmitted via person-to-person contact or through exposure to contaminated food and/or water. The viruses known to be present in relatively large numbers in human feces include cultivable enteroviruses (i.e. echoviruses, coxsackieviruses), adenoviruses, reoviruses, rotaviruses, Hepatitis A virus, and Norwalk-like viruses. Viruses are of particular concern when present in reclaimed wastewater due to their characteristically low (<10) infectious dosages (Murray et al. 2001; Haas et al., 1999).

Viruses evaluated in this project included enteric viruses (enteroviruses and reoviruses) and two types of coliphages. Enteroviruses are a genus of RNA enteric viruses that are small, simply-structured and contain an icosahedral capsid (protein coat). Poliovirus, hepatitis A virus, and many classes of gastroenteritis-causing viruses are examples of enteroviruses. Typically, enteroviruses are associated with domestic wastewater and tend to be more infective and decay more slowly than bacterial pathogens. However, the prevalence and concentration of the hundreds of different enteric viruses in wastewater varies with the health of the community that is served by a wastewater collection system. For example, the recent massive outbreaks of Norovirus, a waterborne pathogen found in sewage, would be associated with increased concentrations of this virus in the wastewater collection system. Because Norovirus is not an enterovirus and is not typically monitored in wastewater, protective measures are needed to ensure that it is removed through wastewater treatment and disinfection.

Table 1-2. Comparison of Characteristics of Indicator Organisms and Pathogens Tested in this Study.

Indicator /Pathogen	Example species or description	Cell wall	Shape	Size, µm	Comments
Bacterial indicators					
Total coliform	*Escherichia, Klebsiella. Citrobacter, Enterobacter*	Gram negative; non-spore forming	Rod	0.5 to 2	Facultatively anaerobic
Fecal coliform	*Escherichia, Klebsiella*; coliforms able to grow at 44.5° C	Gram negative; non-spore forming	Rod	0.5 to 2	Facultatively anaerobic
Enterococci	*Enterococcus faecalis, Enterococcus faecium*	Gram positive, non-spore forming	Cocci	0.5 to 1	Aerotolerant anaerobe
Clostridium perfringens	Opportunistic pathogen; produces enterotoxin	Gram positive, spore forming	Rod	0.6 to 1.3 by 2.4 to 19	Obligately anaerobic
Coliphages	Viruses that infect *E. coli* and other coliform bacteria			0.025– 0.20	
Coliphages that infect *E. coli* F_{amp} host 700891	Male specific (F+) RNA coliphages; Can only replicate when bacterial host cell is in logarithmic growth phase at > 30 °C	No cell wall; coat protein protects RNA	Icosahedral protein shell	0.025	Infect host by attaching to fertility fimbriae
Coliphages that infect *E. coli* 15597 host	Male specific (F+) and somatic coliphages that infect *E.Coli* 15597 (ATTC[a])	No cell wall; coat protein protects nucleic acid	Icosahedral protein shell	0.025	Somatic coliphages attach to cell wall; F+ attach to fertility fimbrae
Enteroviruses	Genus within the family Picornaviridae includes poliovirus, coxsackievirus, echovirus, hepatitis A virus	No cell wall; Non-enveloped protein coat	Icosahedral capsid single strand (ss) RNA genome	0.025-0.030	Infect mammalian cells
Protozoan Parasites	Complex life cycle. Zoonotic (animal to human) transmission.				
Giardia intestinalis	Flagellated protozoan; Phylum Mastigophora		Ovoid cyst	~8.5 x 10	Cyst is infective form
Cryptosporidium parvum	Coccidian protozoan; Phylum Apicomplexa		Ovoid Oocysts;	4 to 6	Oocyst is infective form; resistant to disinfection

[a]ATTC: American Tissue Culture Collection

Coliphages are viruses that infect and replicate in coliform bacteria, thus they can only proliferate when the host bacterium is present. Coliphages can be classified in terms of whether their genome is composed of DNA or RNA. In addition they can be classified as either male-specific (MS) or somatic. MS coliphages require a host that produces the fertility fimbriae (F+). Conversely, somatic coliphages bind to receptors located on the host cell wall and are not restricted to F+ hosts. Because bacteria in biological treatment systems are not in a logarithmic growth phase, it is unlikely that MS coliphages can replicate during treatment.

Similarities in the physical structure, morphology, and nucleic acid composition of certain coliphages and human enteric viruses suggest that reduction of coliphages and enteric viruses may follow similar patterns, depending on the dominant physical, chemical, and/or biological removal mechanisms. In addition, the presence and survival of coliphages is related to bacterial concentrations. Coliphage testing has been proposed as a surrogate indicator for pathogens in wastewater, due in part to the failure of coliform bacteria to correlate with pathogens in wastewater (Havelaar et al., 1993). Two *E. coli* hosts were used in this study to quantify different subsets of coliphages: *E. coli* ATTC strain 700891 (F_{amp}) for male-specific RNA coliphages and *E.coli* ATCC strain 15597 for somatic and male-specific coliphages.

1.2.3 Protozoan Pathogens

Giardia is the most commonly isolated intestinal parasite in the world (Gardner et al. 2001). *Giardia* cysts are present in high numbers in domestic sewage and are of particular concern due to their inherent resistance to disinfectants commonly used in wastewater treatment processes (Rose et al., 1996).

Cryptosporidium is an intestinal parasite found worldwide. The oocysts have been detected in untreated wastewater and also in some drinking water sources (Smith and Rose, 1998; Rose et al., 2002). As with *Giardia*, *Cryptosporidium* oocysts are resistant to the levels of chlorine commonly used in wastewater treatment. *Cryptosporidium* caused the largest waterborne outbreak ever documented in the U.S., where over 400,000 people became ill in Milwaukee, WI when a drinking water treatment plant malfunctioned (MacKenzie et al., 1994). To date, no confirmed illnesses from exposure to reclaimed water sources of *Cryptosporidium* have been documented.

1.3 Regulatory Requirements for Control of Pathogens in Reclaimed Water

Within the U.S., there are no national standards for control of reclaimed water. Therefore, regulatory requirements for control of pathogens in reclaimed water are set by state and local regulatory agencies. A review of regulatory requirements is currently in progress by the U.S. EPA (Crook, 2003). Typically regulatory requirements are linked to the potential uses of the reclaimed water and the potential risks associated with exposure. In some states, the regulations explicitly specify the type of treatment and associated monitoring for individual reuse applications. In other states, the requirements are more generic. The wastewater treatment facilities that participated in this project were from three states: Arizona, California, and Florida. A comparison of the regulatory requirements for these three states is given in Table 1-3.

Table 1-3. Comparison of Microbiological Monitoring Requirements for Arizona, California, and Florida for the Use of Reclaimed Water for Urban Applications.[1]

Parameter	Arizona[2]	California[3]	Florida[4]
Microbial monitoring requirement	Fecal Coliform, ND (7 day median value) 23/100 mL maximum (Class A[5])	Total Coliform; $\leq$ 2.2/100 mL (7 day median) 23/100 mL maximum value in a 30 day period Never exceed 240/100 mL	Fecal coliform, ND in at least 75% of samples Never exceed 25/100 mL
Frequency	Not specified	Daily; compliance is 7 day median value	Daily
Limits	Turbidity < 2NTU 24 hour average; Never exceed 5 NTU	Turbidity < 2NTU; daily average; Cannot exceed 5NTU more than 5% of time; Never exceed 10 NTU	TSS < 5 mg/L $CBOD_5$ < 20 mg/L
Other monitoring requirements	Filtered effluent turbidity	CT[6] 450 mg-min/L; modal contact time of 90 minutes; or 5 log reduction of MS-2 or poliovirus	Minimum chlorine residual 1 mg/L as Cl_2 after 15 min contact time Periodic testing of effluent *Giardia* and *Cryptosporidium* (one sample per 2 years or 5 years depending on plant size)
Treatment requirements			
Biological Treatment	Yes	Yes	yes
Coagulation	Not required; require chemical feed facilities for coagulant and/or polymer addition in case of filter turbidities over 5 NTU (2 NTU 24 hour average)	Needed if secondary effluent turbidity is > 5NTU for a 15 minute period or ever > 10 NTU	Need chemical feed facilities upstream of filtration in case of poor quality secondary effluent
Filtration	Yes	Yes	yes
Disinfection	Yes	Yes	yes

[1]Adapted from Crook, 2003

[2]State of Arizona. 2001. *Regulations for the Reuse of Wastewater.* Arizona Administrative Code, Chapter 9, Article 7, Arizona Department of Environmental Quality, Phoenix, Arizona.

[3]State of California. 2000. *Water Recycling Criteria.* Title 22, Division 4, Chapter 3, California Code of Regulations. California Department of Health Services, Drinking Water Program, Sacramento, California.

[4]Florida Department of Environmental Protection. 1999. *Reuse of Reclaimed Water and Land Application.* Chapter 62-610, Florida Administrative Code. Florida Department of Environmental Protection, Tallahassee, Florida.

[5]Class A includes open access landscape irrigation (parks, residential, schools), recreactional impoundments, food-crop irrigation, closed-loop air conditioning systems, etc.

[6]CT: Product of contact time in minutes and chlorine residual in mg/L

CHAPTER 2.0

METHODS AND MATERIALS

For this project, samples were collected from each of six wastewater treatment facilities over the course of a one year sampling period. At each facility, samples were collected from four locations: influent, secondary effluent, filtered effluent, and disinfected effluent. In each case, grab samples were collected under peak flow conditions. Information on the facilities that participated in the project and sample collection and analysis methods are provided in this chapter.

2.1 Wastewater Reclamation and Reuse Facilities

The six facilities that participated in the project are all full-scale wastewater reclamation and reuse facilities that produce reclaimed water for nonpotable urban applications. Four of the facilities are in Florida, one is in California, and one is in Arizona. A comparison of the six facilities is given in Table 2-1.

The biological process used at four of the facilities is conventional activated sludge. One facility is a nitrification plant and one facility is a biological nutrient removal plant. The types of filters in use included fabric filters, shallow and deep-bed monomedia filters, and dual media filters (sand and anthracite). Of the six facilities, only one routinely pre-chlorinates, one uses a cationic polyelectrolyte, and the others do not routinely use chemicals upstream of filtration. For disinfection, chlorine is used at five of the facilities and one facility uses UV disinfection. For the facilities with conventional activated sludge, the dominant form of chlorine available for disinfection is combined chlorine or chloramines, whereas in the facility with biological nutrient removal, free chlorine is used. The UV system consists of four banks of 64 medium-pressure lamps (Trojan™ 4000). The lamp intensity is based on the UV transmittance and the turbidity of the water.

Table 2-1. Comparison of Facilities Tested During the Project.[1]

Parameter	A	B	C	D	E	F
Range of Capacity MGD	0.9-2.6	13.9-16.2	9.6-10.3	11-25	1.3-2.1	1.3-2.4
Primary treatment	Grit removal	Grit removal	Grit removal Primary clarifier	Grit removal	Grit removal; Equalization basin	Grit removal
Biological Treatment						
Process Type	Activated sludge	Activated sludge	Activated sludge	Activated sludge	Nitrification	Biological nutrient removal
Mean Cell Residence Time, days	6-8	3.5-6	1.6-2.7	3-5	8.7 to 13.3	8 to 16
Filtration						
Filter type	Cloth	Traveling bridge	Monomedium	Dual Media	Continuously Backwashed Upflow Filter	Dual Media
Filter Media (Depth)	Fabric	Sand (0.3 m)	Anthracite (1.2 m)	Anthracite (0.8 m) Sand (0.25 m)	Sand (1.2 m)	Anthracite (0.6 m) Sand (1.2 m)
Chemical Use	None	Pre-chlorinate	Cationic polyelectrolyte	None	None	Alum added to secondary clarifier
Hydraulic Loading Rate gpm/ft^2	1.5 to 5.8	0.6 to 1	1.1 to 1.8	2.2 to 5.8	1.2 to 2	1.3 to 1.5
Volume per backwash, gallons/filter	1,200	3,000	24,000	250,000	8% of flow	60,000
Backwashing frequency, hours	72 to 84	Automatic (every 24)	48	48 to 168	Continuous	48 to 168
Backwash water source	Filtered effluent	Chlorinated effluent (6 mg/L as Cl_2)	Chlorinated Filter effluent (30 mg/L as Cl_2)	Chlorinated effluent (3-6 mg/L); shock chlorinated 3x/year	Final effluent (unchlorinated); shock chlorinated periodically	Chlorinated effluent (5 mg/L as Cl_2)
Disinfection						
Disinfectant	Chlorine or Sodium hypochlorite (since 7-03)	Chlorine	Sodium hypochlorite	Chlorine	Medium pressure UV (256 lamps)	Chlorine
Secondary Effluent Ammonia, mg/L as N	5-12	15-40	11-22	1-5	<0.2	<0.2
Total Chlorine Residual, mg/L	2-5	7-9.5	1.75-4.2	3 - 6	NA	4-6.5
Disinfection pH	6.8-7.2	6.8-7.1	7.2-7.4	7.2-7.4	7.3-7.5	6.9-7.1
Contact time, min	20-80	45-60	340-580	35 -82	NA	70-90
CT, mg-min/L	100-235	350-570	975-2400	120-290	NA	300-570

[1] Based on operating data during sampling

2.2 Sample Collection

Samples were collected from four sites within each plant to represent effluents from each stage of the treatment process. Sampling locations consisted of wastewater influent (site 1), secondary effluent (site 2), filter effluent (site 3), and disinfected effluent (site 4). Each facility was sampled five times; approximately every two months. Samples were either collected by project personnel (Facilities A, B, D, and F) or by operators at the individual facilities (Facilities C and E). Depending on the treatment plant location, samples were either shipped from the treatment plant to the analytical laboratories (Florida, Michigan) or transported back to USF for initial processing and shipping. Each sample was assayed for traditional and alternative bacterial and viral indicators, parasites, and enteric viruses within 32 h of collection (depending on the facility location and type of analysis performed).

The sampling for each plant began during the peak flow for the day of sampling (usually mid-morning). After collecting the influent sample, attempts were made to collect downstream samples based on the approximate hydraulic detention time of each treatment unit. All samples were collected in sterile containers and were stored on ice until they were processed or transferred to a refrigerator. Chlorinated samples (disinfected effluent and filtered effluent from facility that pre-chlorinate) were dechlorinated with sodium thiosulfate added to a final concentration of 100 mg/L. The volume of sample collected at each site varied by parameter and location. For bacterial and coliphage enumeration, sample volumes were 50 mL of influent, 500 mL from the secondary clarifier, 2 L of filtered effluent and 2 L of disinfected effluent. For enteric virus assays, the sample volume was based on either the amount of water that could be processed without clogging the filter or a fixed volume. Typically less than 100 L was filtered for each influent sample, depending on water quality (i.e. suspended solids content). Standardized sample volumes were used for the other sample locations: 190 L samples from the secondary clarifiers, and 378.5 L samples from the filtration and disinfection steps. For quantification of protozoan parasites, the sample volumes used were about 0.5 – 1.0 L influent, about 19 L secondary effluent, about 38 L effluent from filters, and about 53 L disinfected effluent.

Grab samples were used for the bacterial and viral indicators, while the parasite and virus samples were collected and concentrated by pumping large volumes of water through appropriate filters. 1MDS (Cuno, Inc.) cartridge filters were used for collection and concentration of viruses while Pall-Gelman Envirochek HV capsule filters were used for collection and concentration of protozoan parasites.

Samples were placed on ice and shipped to the appropriate laboratories for analysis (University of South Florida, Michigan State University, University of Florida). Operational data associated with each sampling event was provided by operators at each facility. In some cases, supplemental water quality analyses were conducted by the project team.

2.3 Enumeration of Bacterial Indicators and Viral and Protozoan Pathogens

The microbiological assays included total and fecal coliform bacteria, enterococci, coliphages grown on two different *E. coli* hosts, enteroviruses, *Giardia,* and *Cryptosporidium*. Methods are summarized below.

Coliform bacteria Total coliform (TC) were isolated by membrane filtration according to Standard Methods for the Evaluation of Water and Wastewater (APHA, 1998). Samples were filtered through 47 mm cellulose acetate membrane filters with a nominal pore size of 0.45 μm. Tenfold serial dilutions were made up to 1:10,000. One mL of each dilution was added to phosphate buffered saline (PBS), and filtered. A maximum of 100 mL of undiluted sample was filtered. Each volume was filtered in triplicate. Filters were transferred to 50 mm petri dishes containing mEndo-LES agar (Difco Laboratories, Detroit, MI) and incubated for 24 h at 37°C. Colonies that produced a metallic sheen were enumerated as total coliforms. For enumeration of fecal coliform, membrane filters were transferred to 50 mm petri dishes containing mFC medium (Difco) and incubated for 24 h at 44.5°C in a water bath. Blue colonies were enumerated as fecal coliforms. *Escherichia coli* (American Type Culture Collection [ATCC]# 9637) was used as the positive control for all coliform measurements.

***Enterococcus* spp.** were enumerated using a modification of Method 1600 (U.S. EPA, 1997). Samples were filtered as described above and the filters were transferred to 50 mm petri dishes containing mEI agar. This media consisted of mE agar base (Difco) amended with indoxyl β-D glucoside (0.75g/L), nalidixic acid (0.24 g/L) and triphenyl tetrazolium chloride (0.02 g/L). Plates were incubated at 41 ± 0.5°C. After 24 h incubation, colonies exhibiting a blue halo were enumerated as enterococci. *Enterococcus faecalis* (ATCC #19433) was used as a positive control.

C. perfringens were enumerated by membrane filtration and filters were transferred to Petri plates containing mCP agar (Acumedia Manufacturers, Inc. Baltimore, MA). Plates were incubated anaerobically using anaerobic gas packs (BBL GasPak, Beckton Dickinson) at 45°C for 24 hours. The resultant yellow colonies were exposed to ammonium hydroxide fumes and the colonies that turned red or dark pink were enumerated as *C. perfringens* (Bisson and Cabelli, 1979). *C. perfringens* (ATCC #13124) was used as positive control.

Coliphages were analyzed by two methods: the agar overlay method of Adams (1959) and a version of the large volume (1L) presence/absence assay of Yanko et al. (1999) for treated effluent. Two *E. coli* host strains were used in separate assays: *E. coli* HS(pFamp)R (ATCC #700891), which hosts male-specific (F+) coliphages, and *E. coli* C-3000 (ATCC #15597), which should host both somatic and F+ coliphages. Serial dilutions of samples were made in phosphate buffered saline (PBS) buffer according to expected phage concentrations at each treatment step. Five replicate volumes of 0.1 ml to 2 ml were plated for each dilution except in the case of the disinfected effluent samples, for which ten replicates of 2 ml each were plated. Plaque forming units (PFU)/100 mL were calculated after 24 h incubation. In this report

$Coliphage_{700891}$ refers to the male-specific coliphages and $Coliphage_{15597}$ refers to the *E. coli* C-3000 host.

Enteroviruses Samples were filtered through Virusorb 1MDS Filters (Cuno Inc., USA) as per U.S. EPA (1996) methodology. Typically less than 100 L was filtered for each influent sample, depending on water quality (i.e. suspended solids content). Sample volumes were approximately 190 L from the secondary clarifiers, and 378.5 L from the filtration and disinfection steps. Filters were eluted with 1 L of 1.5% beef extract (BBL V) in 0.05 M glycine (pH 9.5, ~25°C) (U.S. EPA/ICR). The eluted sample was concentrated by organic flocculation and assayed for enteric viruses by the observation of cytopathic effects (CPE) on recently passed (<4 days) cell lines. Three cell lines, Buffalo Green Monkey (BGM), Rhabdosarcoma (RD), and MA-104 cells were used for this purpose. Positive controls were performed in a separate room using Poliovirus I. Cytopathic effects on each cell line were observed, and the most dilute sample showing CPE was recorded. Most probable number (MPN) determinations were performed using U.S. EPA released software.

Cryptosporidium* and *Giardia For the detection of parasitic protozoa, samples were filtered through a high-volume Gelman Envirochek HV filter using a portable electric diaphragm water pump with flow rates maintained at 2 L/min. Sample volumes varied depending upon the treatment stage and the amount of water that could be filtered, i.e. 0.5 – 1.0 L influent, ~19 L secondary effluent, ~38 L effluent from filters, and ~53 L disinfected effluent. Detection limits varied with the total volume of sample filtered. Filters were eluted and processed according to Method 1623 (U.S. EPA, 1999) and sample concentrates were subjected to immunomagnetic separation for purification of the oocysts from the material captured. Aliquots of the purified sample concentrate were fixed with methanol onto well slides, filtered through cellulose acetate membrane filters or removed into 1.5-mL microcentrifuge tubes. For *Cryptosporidium* only, each sample was split between cell culture viability testing and microscopic enumeration. Tissue cultures were fixed with 100% methanol for 10 minutes and labeled using the indirect antibody procedure [Foci Detection Method (FDM-MPN)] (Slifko, et al., 1997, 1999). Samples were examined under by both epifluorescent and DIC microscopy. Equivalent volumes were calculated, and results are reported as cysts or oocysts per 100 L. Method 1623 determines the total numbers of IFA positive oocysts and cysts (empty, full, potentially viable) and does not address the viability of the oocysts or cysts.

Stock solutions of 4',6-diamidino-2-phenylindole (DAPI, [catalog no. D-9542; Sigma Chemical Co., St. Louis, MO] 2 mg/mL DAPI in absolute methanol) and propidium iodide (PI, [catalog no. P-4170; Sigma Chemical Co., St. Louis, Mo.] 1 mg/mL PI in 0.1 M PBS, pH 7.2) were prepared and stored at 4 °C in the dark. From the DAPI stock solution, a 1:5,000 dilution in PBS was prepared. Each subsample was stained as follows: (i) fluorescein-labeled antibody (EasyStain, Biotechnology Frontiers, Australia and Giardi-a-Glo™, Waterborne™, Inc. New Orleans, LA) and DAPI; (ii) fluorescein-labeled antibody and PI; (iii) fluorescein-labeled antibody and DAPI/PI.

The IF staining and dye uptake protocols were performed in well slides (DYNAL spot-on, DYNAL, Inc. New York) and on 0.22 µm pore size Sartorius 25-mm diameter cellulose acetate filters (Sartorius Corp., Edgewood, NY) following the manufacturer's specifications. After the incubation period, each subsample was stained with DAPI, PI, or a combination of DAPI and PI. One-hundred µL and 500 µL of each fluorochrome were added to methanol-fixed and membrane-filtered subsamples, respectively. Some of the samples fixed on well slides received 10 µL of each fluorochrome stock solution according to specifications by Campbell et al. (1992) for *Cryptosporidium* oocysts. DAPI-stained sub-samples were incubated for 2 minutes USEPA Method 1623; U.S. EPA, 1999) while PI-stained subsamples were incubated for periods of 2 and 10 minutes; (Sauch et al. 1991). DAPI/PI stained subsamples were incubated for periods of 2 and 10 minutes at room temperature and 2 hours at 37°C, which is the optimum time for maximum dye uptake for *Cryptosporidium* oocysts (Campbell et al., 1992). Two of the subsamples were stained in suspension with the FITC-labeled antibodies after the DAPI/PI exposure step, which was performed at 37°C for 2 hours. Some of the membrane filters were dehydrated by sequential application of 20, 40, 80, and 90.2% ethanol solutions containing 5% glycerol.

Microscopic examination Samples were observed with the 20X and 40X Dplan Apo UV objective lens with both epifluorescence and DIC microscopy. An Olympus model BH2 microscope equipped with a UV filter block (350-nm excitation and 450-nm emission) for DAPI, blue filter block (450-nm excitation and 520-nm emission) for FITC and a green filter block (500-nm excitation and 630-nm emission) for PI was used. Nomarski differential interference contrast microscopy (DIC) was used to look at internal morphological characteristics in *Giardia* cysts under 400X magnification.

***Cryptosporidium* viability assay** Concentrates from the IMS procedure were inoculated onto HCT-8 cell monolayers in eight-well chamber glass cell culture slides as described by Slifko et al (1997). RPMI media was used and cells were plated in eight-well chamber glass cell culture slides. Water samples were treated with 0.525% sodium hypochlorite (10% reagent grade bleach (vol/vol) in DI water) at 4°C for eight minutes and washed once by centrifugation, for the purpose of surface sterilization. The bleach treatment triggers the oocysts to excyst and prevents the tissue cultures from becoming contaminated by bacteria, fungi or viruses present in the oocyst samples. The washing serves to remove residual chlorine to prevent interference with cell culture. Samples were pipetted onto cell monolayers in six replicate wells so that the entire volume of each concentrate was inoculated. For each well slide, one well of uninoculated cell monolayer was set up concurrently as a negative control and as a control of the health of the cell monolayer, and a well containing bleach-treated fresh oocysts (less than 30 days old) was used as a positive control. The cultures were incubated in a 5% CO_2 atmosphere at 37°C for 48 hours. The presence of multiple stages of the parasite and clustering (indicating the auto-reinfective nature of the life cycle) were recorded. Infectious *Cryptosporidium* were enumerated by the Foci Detection-Most Probable Number Method described by Slifko et al. (1999). Results were reported as viable oocysts per 100L.

2.4 Data Interpretation and Statistical Analysis

All microbiological data were log_{10} transformed and statistical analysis was performed to assess inter- and intra-facility differences in levels of indicator and pathogenic organisms due to different treatment design and operational variables. The design variables and treatment processes evaluated included: biological treatment parameters, type of filtration, and method of disinfection. Operational variables evaluated included solids retention time, filter loading rates, chemical usage, backwash practices, disinfectant residuals, and contact times. Trends of pathogen reduction and survival were compared to operating parameters for each process for each plant individually and for pooled data from all plants to identify potential correlations.

Statistical analyses were performed using Microsoft Excel, SAS™, Kaleidograph™, and ProStat™ software. In cases where microbial concentrations were below detection limits, such as for samples from the final disinfected effluent, the data presented used the detection limit to calculate the mean. In some cases, additional analyses were conducted using 0-values or ½ the detection limit; however, these calculations did not affect the results of the statistical analysis and are specifically mentioned where they occur. Parametric statistics were used for normally distributed data (most analyses); exceptions are noted. Post-hoc comparisons between treatments utilized the Tukey-Kramer multiple comparisons test. Linear correlations were conducted on normally distributed data using Pearson r; r^2 is reported. Differences in frequency of detection of microbial analytes were assessed by Chi-square analysis.

CHAPTER 3.0

RESULTS

This chapter will summarize the concentrations of indicator and pathogenic microorganisms in the untreated, secondary effluent, filtered effluent and disinfected final reclaimed water from six facilities. There are a number of approaches for addressing water quality in the wastewater and reclaimed water arena however the focus is generally on final concentrations in the discharged effluent from the facility. There has been very little focus on defining treatment removal goals (eg. should filtration in a reclamation facility achieve 90, 99, or 99.9% reduction of protozoa?). Thus while removal by the individual processes is of interest and will be presented, the major discussion will be centered on the microbial water quality observed during this study. Appendix A includes a summary table of arithmetic averages (detection limits used to calculate these averages in samples where the constituent was not detected), numbers of samples positive and collected as well as percent removal from secondary, filtration and disinfection processes at each facility.

3.1 Concentrations of Indicators and Pathogens in Untreated Wastewater

The microbial quality of the wastewater influent was fairly consistent among the treatment facilities. All traditional and alternative indicators as well as viral and protozoan pathogens were above detection limits in all influent samples collected with the exception of *Clostridium perfringens* (detected in 93% of the samples) *Cryptosporidium* (detected in 74% of influent samples) and viruses (detected in 93% of the samples). A comparison of bacterial indicator concentrations in the influent to each facility is shown in Figure 3-1 in a boxplot format. The box represents 50% of the data values and the horizontal line signifies the median value associated with each facility; the 95% confidence interval is shown by the lines that extend above and below the boxes and outliers are represented by circles. For all plants, total coliform levels ranged from 10^6 to 10^7 cfu per 100 mL and fecal coliform levels ranged from 10^5 to 10^7 cfu per 100 mL. Enterococci concentrations ranged from 10^4 to 10^6 cfu per 100 mL. Concentrations of *Clostridium perfringens* were more variable with concentrations ranging from 10^2 to about 10^6per 100 mL. There was not a statistically significant difference in the concentrations of any of the bacterial indicators among the six treatment plants, with the exception of *Clostridium perfringens*, which was slightly higher on average in Facility C.

A comparison of the concentrations of coliphages in the influent to each of the six wastewater treatment facilities is shown in Figure 3-2. Concentrations of coliphage$_{15597}$ ranged from 10^3 to 10^6 pfu/100mL. The concentrations of coliphage$_{700891}$ (f-amp) were more variable within each plant and ranged from 10^2 to 10^8 pfu/100 mL.

A comparison of the concentrations of enteric viruses isolated from the untreated wastewater is shown in Figure 3-3. Concentrations of enteric viruses (MPN) ranged from about 10^2 to 10^4 MPN/100 L.

A comparison of the concentrations of protozoan pathogens (total IFA oocyst and cyst counts) in the untreated wastewater is shown in Figure 3-4. In general, concentrations of *Giardia* in the untreated wastewater sources were about an order of magnitude higher than concentrations of *Cryptosporidium* for each of the facilities tested. *Giardia* concentrations ranged from about 10^2 to 10^6 cysts/100 L and *Cryptosporidium* concentrations ranged from 10^1 to 10^4 oocysts/100 L.

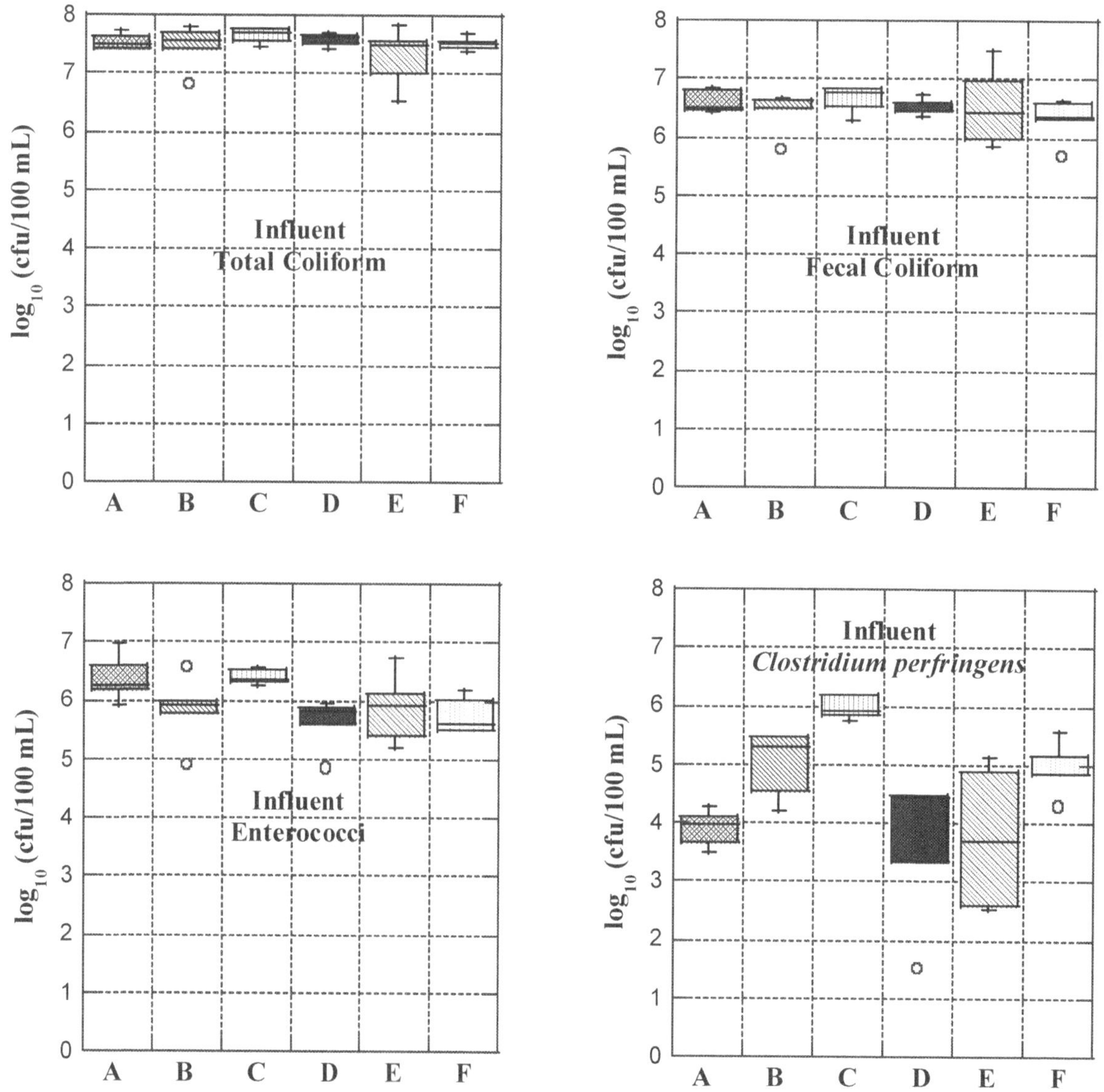

Figure 3-1. Boxplot Comparison of the Concentrations of Bacterial Indicators Isolated from the Influent to Each of Six Wastewater Treatment Facilities (A-F). Facility Descriptions are Given in Table 2-1.

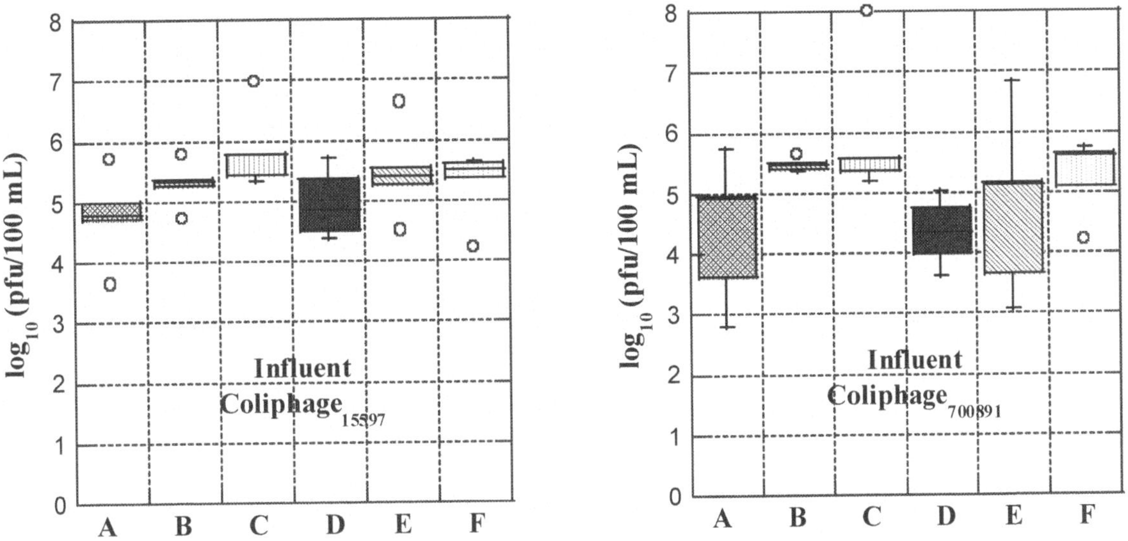

Figure 3-2. Comparison of the Concentrations of Coliphages Isolated from Untreated Wastewater from Each of Six Wastewater Treatment Facilities. Facility Descriptions are Given in Table 2-1.

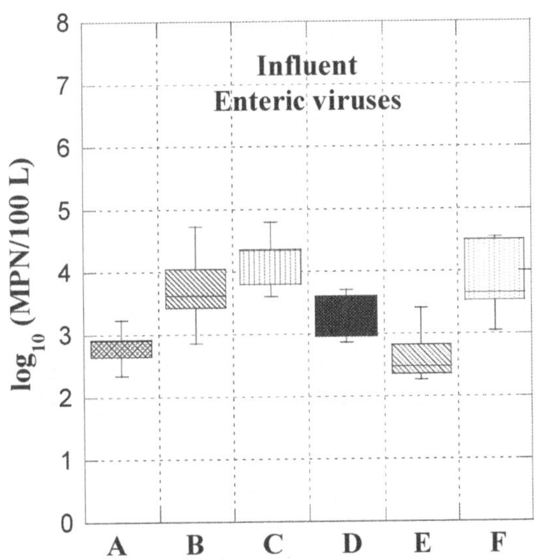

Figure 3-3. Comparison of Concentrations of Enteric Viruses Isolated from Untreated Wastewater from Each of Six Wastewater Treatment Facilities (A-F), Facility Descriptions are Given in Table 2-1.

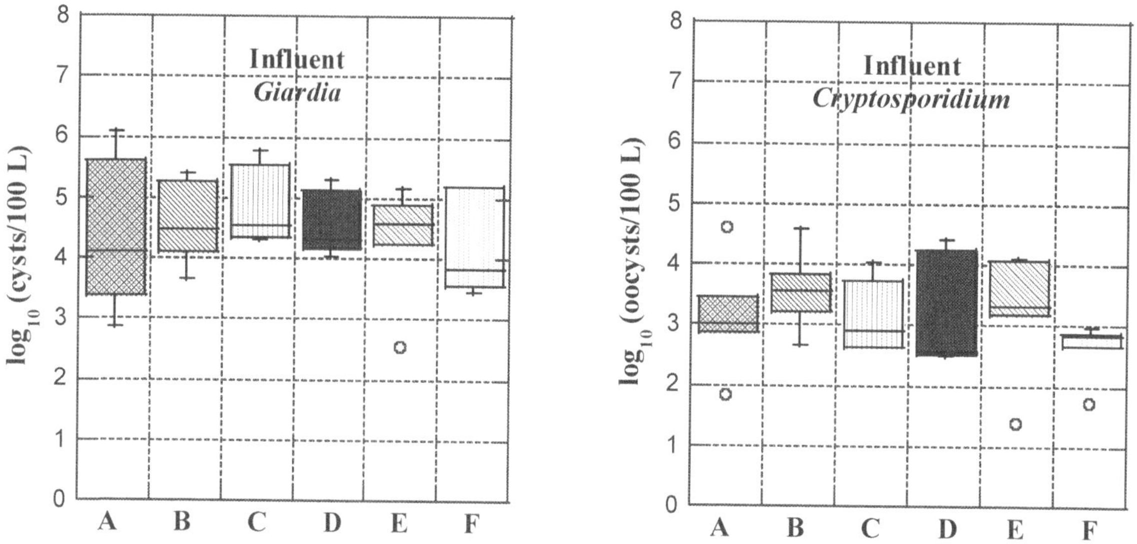

Figure 3-4.. Comparison of Concentrations of *Giardia* Cysts and *Cryptosporidium* Oocysts in Untreated Wastewater from Each of Six Wastewater Reclamation Facilities. Facility Descriptions are Given in Table 2-1.

3.2 Reduction of Indicators and Pathogens Through Biological Treatment

The biological treatment units in operation at the facilities tested in the project included conventional activated sludge (Facilities A-D), biological nitrification-denitrification (Facility E), and biological nutrient removal (Facility F). A comparison of the concentrations of indicators and pathogens in secondary effluents is given in this section. Key process variables that impact removal of indicators and pathogens are discussed.

In general, the removals by secondary treatment for bacteria and viruses ranged from 96-99.9%, with *Clostridium* and the somatic phage removals ranging from 93.3-99.8% and 13-99.9%, respectively. In all the facilities the highest removal of bacteria and viruses was achieved in Facility F, and the lowest removal in Facility C followed by A. The parasite total cysts and oocysts were removed by secondary treatment the least by Facilities D and F, respectively and the best by Facility E (ranges of 97.7-99.8% removals and 0-99.4% removals for cysts and oocysts, respectively).

3.2.1 Concentrations of Indicators and Pathogens in Secondary Effluents

A comparison of the concentrations of bacterial indicators isolated from secondary effluents from the six facilities tested in this study is shown in Figure 3-5. The concentrations of total coliform in the secondary effluents from each plant ranged from 10^2 to 10^6 cfu/100 mL. Fecal coliform concentrations ranged from 10^1 to 10^5 with the lowest concentrations associated with the biological nutrient removal facility (F). Facility D had the most variability in concentrations of indicators in the secondary effluent, perhaps due to contributions from runoff during the sampling events. Enterococci levels were similar to fecal coliform levels and median values ranged from about 10^3 cfu/100 mL (biological nutrient removal) to 10^5 (conventional activated sludge). *Clostridium perfringens* levels were more variable among the facilities and within each facility with concentrations ranging from 10 to 10^4 cfu/100 mL.

A comparison of the concentrations of coliphages isolated from secondary effluents is shown in Figure 3-6. The concentrations of the $coliphage_{15597}$ host (somatic and MS) ranged from about 10 to 10^5 pfu/100 mL and were more variable than the concentrations of the $coliphage_{700891}$ f-amp (MS). For the four activated sludge facilities (A,B,C,D), coliphage levels were lowest in the secondary effluents from plant F, which has a longer MCRT.

A comparison of the concentrations of enteric viruses isolated from secondary effluents is shown in Figure 3-7. The concentrations of viruses ranged from below 10 to about 10^2 MPN/100 L with the highest concentrations of enteric viruses associated with plant C followed by plant B. Enteric virus detection limits for the secondary effluent samples ranged from 2.9 to 11 MPN/100 L, depending on the sample volume processed. The concentrations of enteric viruses in about 27% of the secondary effluent samples were below detection limits, with nondetects associated with all facilities except plant C (shortest MCRT).

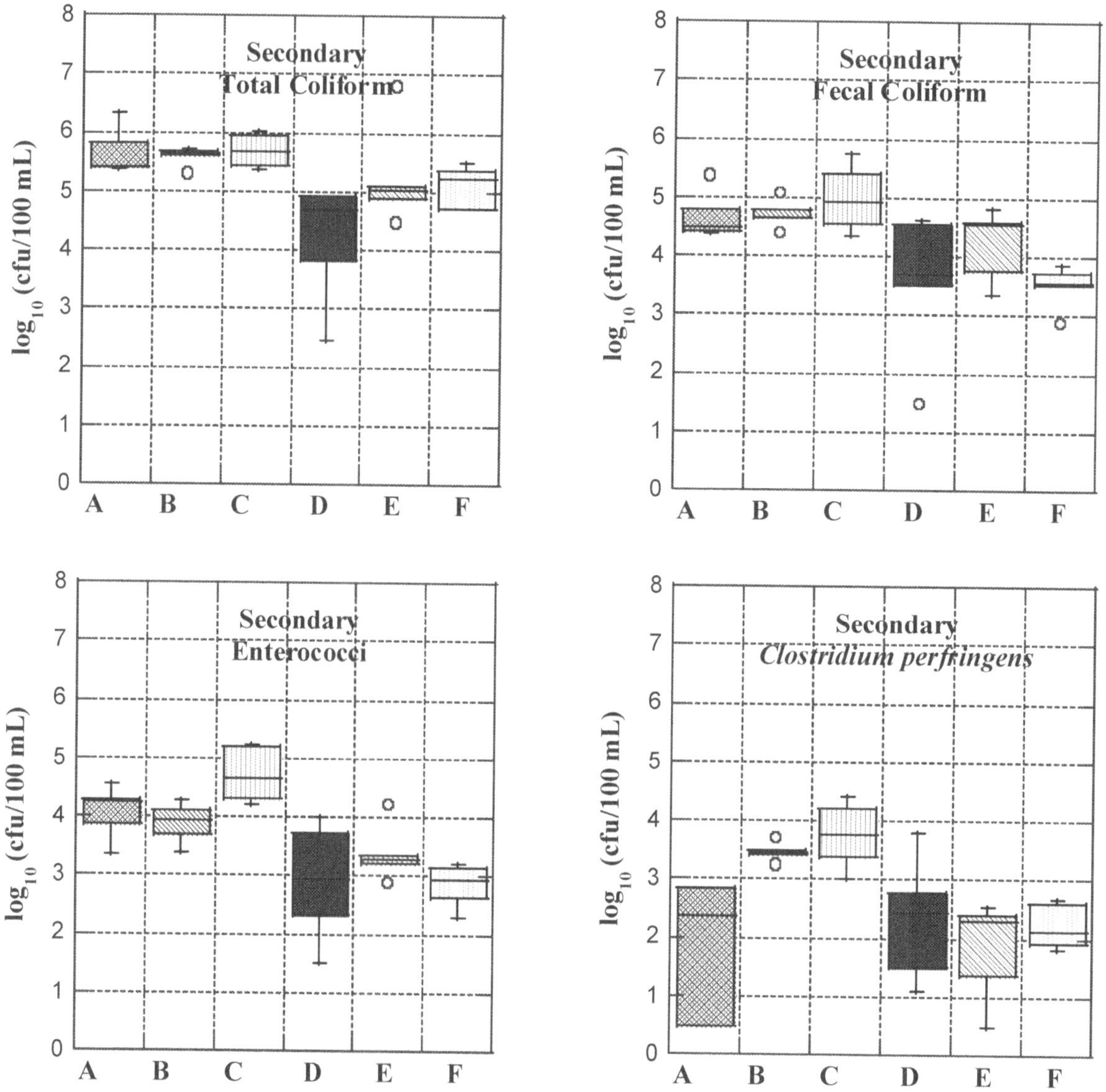

Figure 3-5. Comparison of Concentrations of Bacterial Indicators Isolated from Secondary Effluents from Each of Six Wastewater Reclamation Facilities. Facility Descriptions are Given in Table 2-1.

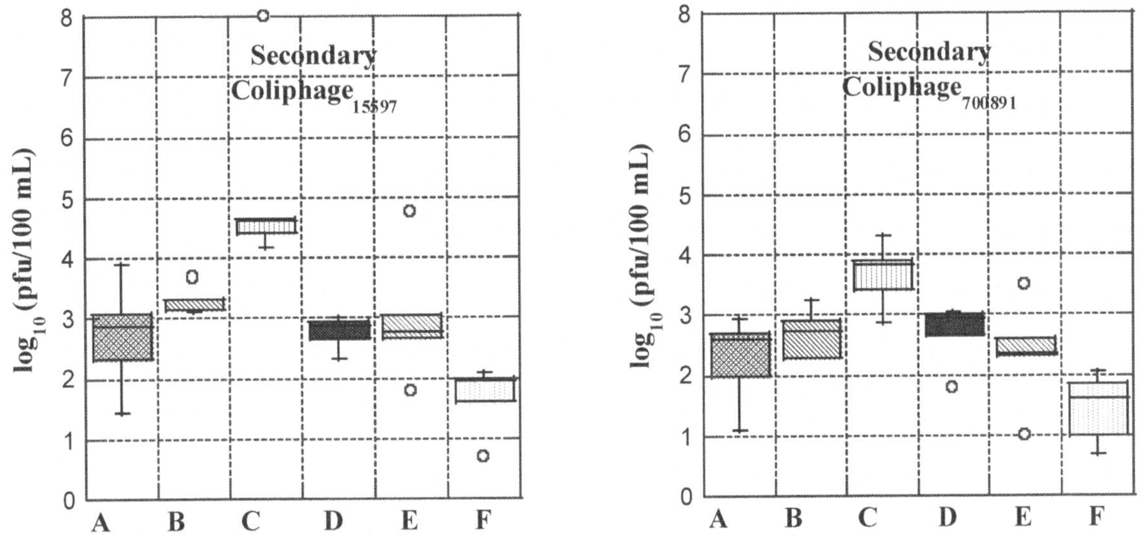

Figure 3-6. Comparison of Concentrations of Coliphages Isolated from Secondary Effluents from Each of Six Wastewater Reclamation Facilities. Facility Descriptions are Given in Table 2-1.

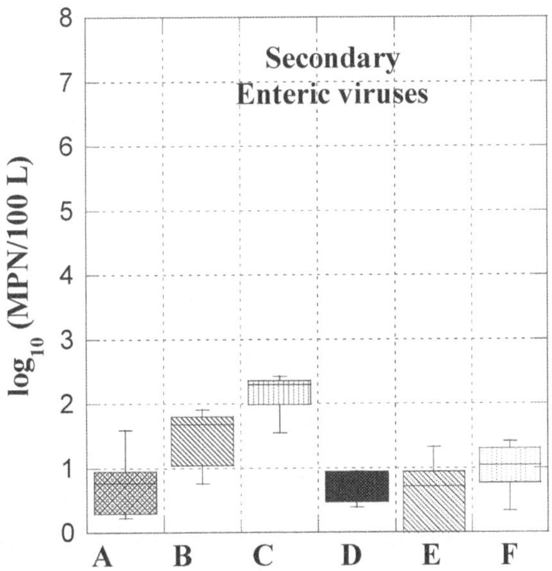

Figure 3-7. Comparison of Concentrations of Enteric Viruses Isolated from Secondary Effluents from Each of Six Wastewater Reclamation Facilities. Facility Descriptions are Given in Table 2-1.

A comparison of the concentrations of *Giardia* and *Cryptosporidium* isolated from secondary effluents is shown in Figure 3-8. The values reported are total numbers of cysts or oocysts and do not provide information on pathogen infectivity. Concentrations of *Giardia* ranged from below 10 to about 10^4 cysts/100 L with the lowest concentrations associated with plant D. Concentrations of *Cryptosporidium* ranged from about 10 to 10^3 oocysts/100 L and were similar for all facilities. For *Giardia* 6% of the samples were below detection limits of 19 cysts/100 L (Plants A and E). For the *Cryptosporidium* analyses, 16% of the samples were below detection limits which ranged from 13 to 94 oocysts/100 L. The nondetects were associated with secondary effluents from plants C, E, and F.

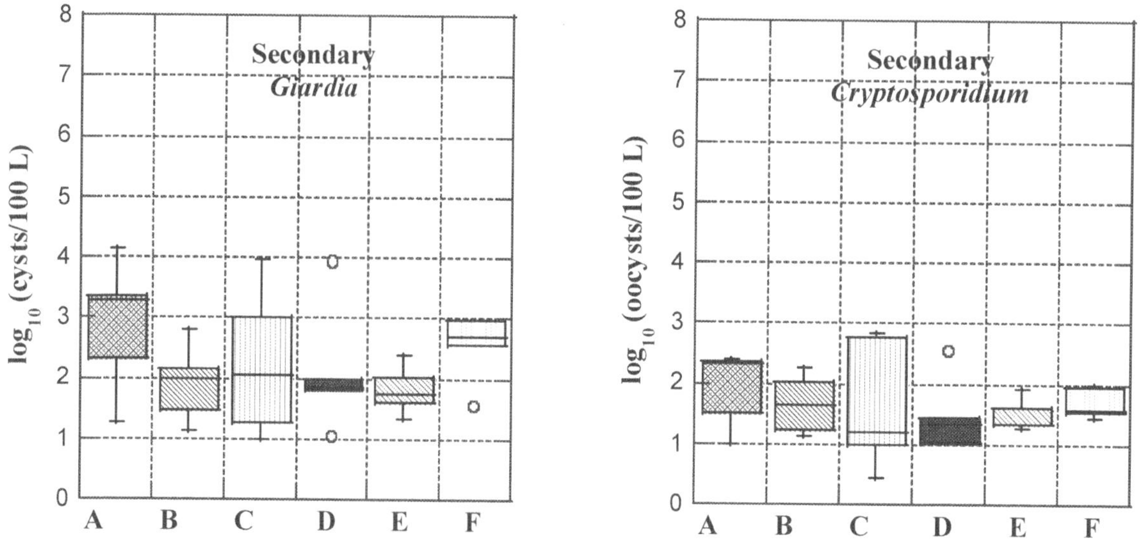

Figure 3-8. Comparison of Concentrations of Protozoan Pathogens Isolated from Secondary Effluents from Each of Six Wastewater Reclamation Facilities. Facility Descriptions are Given in Table 2-1.

3.2.2 Evaluation of Process Variables

The biological processes associated with the facilities tested for this project varied in terms of MCRT, MLSS, the degree to which nitrification occurred, the extent of nutrient removal (nitrogen and phosphorus), and the addition of chemicals to the secondary clarifier. The indicator and pathogen data were evaluated in the context of process variables to identify factors that influence removal through secondary treatment. A comparison of the concentrations of fecal coliforms, enterococci, and enteric viruses in secondary effluent as a function of MCRT is shown in Figure 3-9. In general, similar removal patterns were observed for fecal coliforms, enterococci, and enteric viruses. Correlation coefficients (r^2) for log-transformed concentrations of microorganisms vs. MCRT were ~0.5. Organism concentrations decreased with increasing MCRT by a factor of 0.13 (fecal coliforms and enterococci) to 0.17 (enteric viruses) times the MCRT. For the other biological indicators and pathogens tested in this project, removal or secondary effluent concentrations did not correlate to MCRTs.

A comparison of the secondary effluent concentrations of fecal coliforms, enterococci, and coliphage$_{15597}$ host as a function of MLSS is shown in Figure 3-10. In general, concentrations of these indicators decreased with increasing MLSS, perhaps due to the increased potential for entrapment in the biological flocs. The correlation coefficient, r^2, was 0.5 for the enterococci and 0.6 for the fecal coliforms and coliphages. The concentration of the bacterial indicators decreased at a rate of -0.006 cfu/100 mL per mg/L of MLSS. The decrease in coliphage concentration was about -0.001 pfu/100 mL per mg/L of MLSS.

WERF

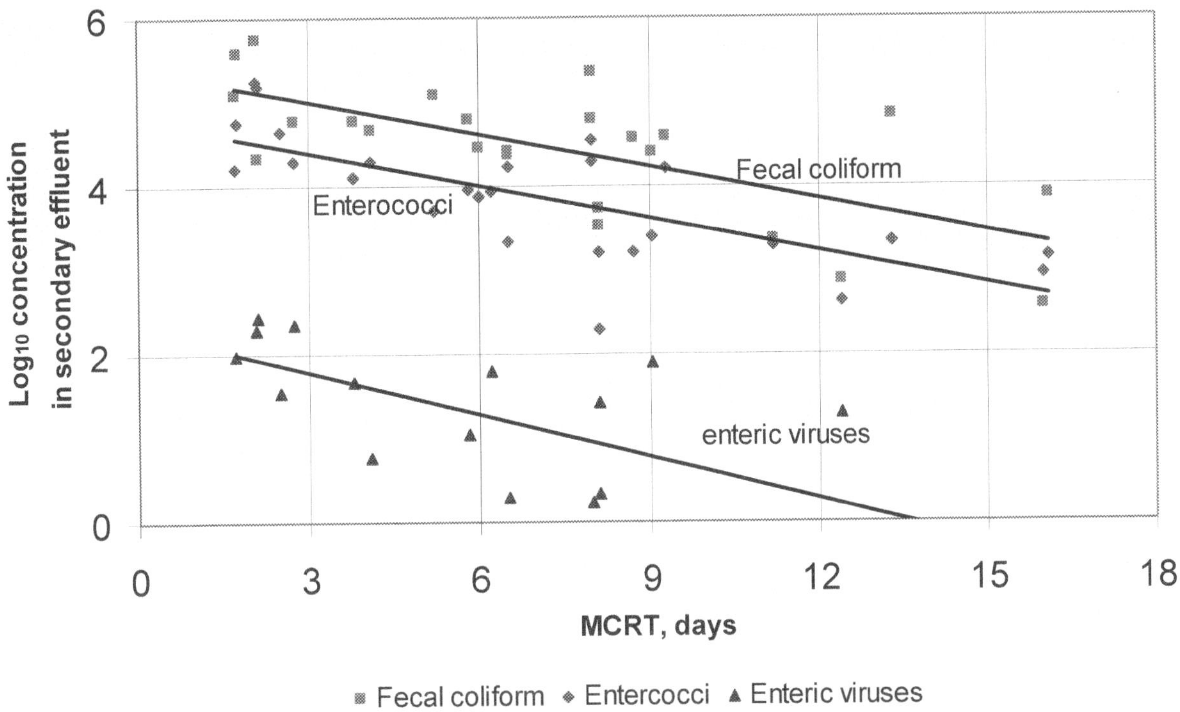

Figure 3-9. Secondary Effluent Concentrations of Fecal Coliforms (cfu/100mL), Enterococci (cfu/100 mL), and Enteric Viruses (MPN/100L) as a Function of MCRT. Corelation Coefficients, r^2, for Each Parameter are 0.5.

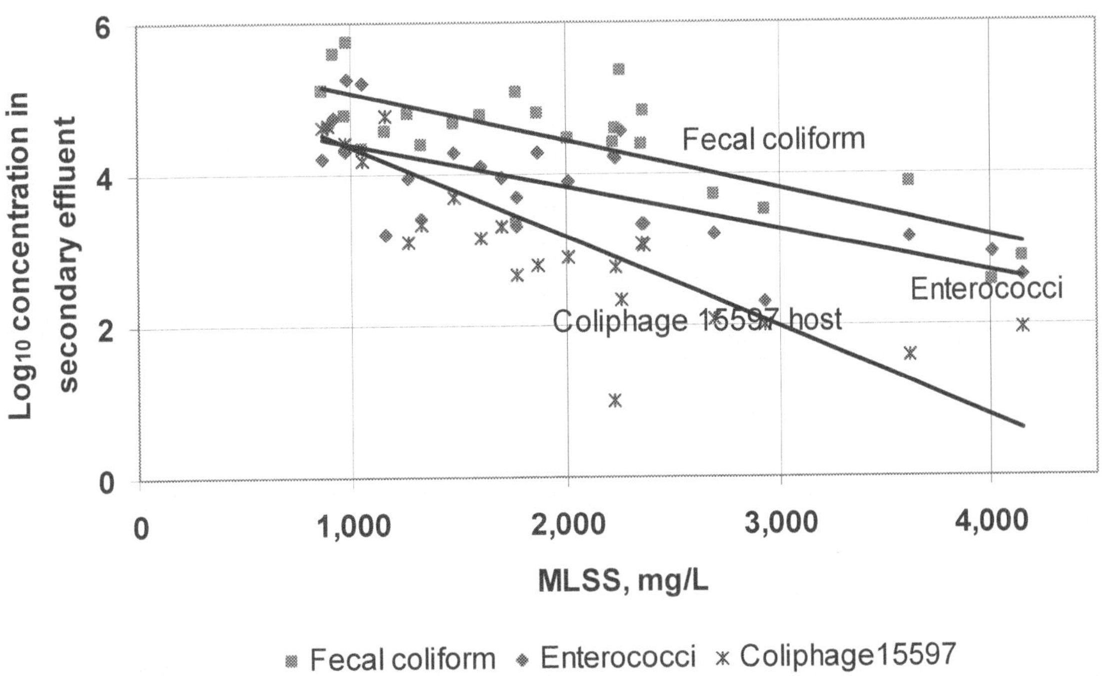

Figure 3-10. Secondary Effluent Concentrations of Fecal Coliforms (cfu/100mL), Enterococci (cfu/100 mL), and Coliphage15597 Host (pfu/100 mL) as a Function of MLSS. Corelation Coefficients, r^2, are 0.5 (Enterococci) and 0.6 (Fecal Coliforms and Coliphage).

3.3 Reduction of Indicators and Pathogens Through Filtration

The types of filters in use at each facility include cloth filters (Facility A), shallow-bed sand filters (Facility B), and deep bed filters (mono-media anthracite or sand, and dual media (anthracite and sand) (Facilities F, E, C, and D). The concentrations of indicators and pathogens in the effluents from the filters are summarized in this section. The impact of process variables including hydraulic loading rate and filter depth on reduction of indicators and pathogens is discussed.

There has always been a question of whether upstream concentrations of microorganisms impact the efficiency of a process or is it primarily influenced by the design and operation. Removals at each facility from the average entering the filtration process for each microorganism were examined (Appendix A). In this study, there was no indication that those facilities with the greatest concentrations coming into the filtration process from secondary had a greater or lower removal. Thus design and operations were the predominant factors as expected. Removals by the various filters and operations employed were variable. The bacteria were reduced between 74-99.9%, with the exception of fecal coliform bacteria which were reduced in concentrations in the range from 33-99.6% by the six facilities. No specific trend was seen, as no one facility provided the greatest removals and another the lowest. In Facility B, chlorine plus filtration did not show on average the greatest removal compared to the other facilities. However, it appears from the specialized studies that Facility B filters were able to equal the reduction via disinfection and inactivation processes of bacteria and viruses because of the predisinfection. When a disinfectant is applied prior to filtration, thoseorganism sensitive to the disinfectant will be inactivated to some degree, which is reflected as removal by the filter. Inactivation, however, is no the same mechanism as physical removal or retention of particles by the filter.

Coliphage and viruses were removed by filtration at a lower rate, from 0-99.2%, 0-82.9% and 46-95.1%, and the parasite removal was also lower, ranging from 61.3-98.9% and 0-99.4% for cysts and oocysts reductions, respectively. However in these cases there was a clearer trend with depth filters (at Facilities F, followed by C with added coagulation, E and finally D, with decreasing order of filter depth) providing the highest average removals. Facility B with chlorination impacted coliphage and virus reductions via inactivation and finally Facility A most often provided the lowest reductions.

3.3.1 Concentrations of Indicators and Pathogens in Filtration Effluents

A comparison of the concentrations of bacterial indicators in filtered secondary effluents is shown in Figure 3-11. Total coliform concentrations ranged from about 1 to 10^5 cfu/100 mL, with detection limits varying from 0.4 to 0.6 cfu/100 mL. Total coliform levels in about 6% of the filtration effluent samples were below detection limits. Mean (log_{10}-transformed) values were significantly different by facility (P=0.002); total coliform concentrations at Facility B (which practices prechlorination) were significantly lower than those at Facilities C, D, and E (nondetects were represented by the detection limit for all of the ANOVA calculations). Similar trends were observed for fecal coliforms and enterococci. Mean log_{10}-transformed fecal coliform concentrations were significantly different by facility (P=0.009), and fecal coliform concentrations at Facility B were significantly different from those at Facilities C, D, and E. Logtransormed *Enterococcus* means were significantly different by facility (P=0.0018), and the mean value at Facility B (due to inactivation/prechlorination) was significantly lower than those at facilities C and D. Detection limits for fecal coliforms and enterococci ranged from 0.3 to 333 cfu/100 mL. About 35% of the fecal coliform samples and 16% of the enterococci samples were below detection limits. The nondetects were associated with samples from plants A, B, C, and F. Slightly different patterns were observed for the other bacterial indicator, *Clostridium perfringens*. Mean concentrations were also significantly different by plant for *C. perfringens* (P=0.0217), but the values that were significantly different were A vs. C and C vs. D. In this case, 21% of the filtration effluent samples were below detection limits (ranging from 0.4 to 333 cfu/100 mL) and the concentration ranged from about 1 to 100 cfu/100 mL in the filtration effluents from all plants except plant C.

A comparison of the concentrations of coliphages isolated from filtration effluents from each facility is shown in Figure 3-12. About 20% of the filtered effluent samples were below detection limits (5 to 10 pfu/100 mL) with most of the nondetects associated with plants A, D, and F. The concentrations of coliphages ranged from about 10 to 10^4 cfu/100 mL, and $\log_{10}$-transformed means were significantly different by facility (P=0.013). The highest mean concentration was associated with plant C (monomedia anthracite declining rate filter), which was significantly greater than values at Facilities D and F. Filtration Facility C achieved on average 99.2% and 50.6% removals of the somatic and f-amp coliphage, respectively. Facility C had the highest concentrations in the secondary effluent entering the filtration process. Prechlorination may have had some minimal impact on the reductions of coliphage by filtration but it is difficult to discern as the average reductions were for Facility B were 89% and 61%. This impact was clearer for the enteric viruses (see below)

A comparison of the concentrations of enteric viruses in the filtration effluents from the six wastewater reclamation facilities is shown in Figure 3-13. Virus levels ranged from about 1 to 10^2 MPN/ 100 L, and mean concentrations ($\log_{10}$-transformed) were significantly different by facility (P=0.0001). The mean enteric virus concentration at Facility C was significantly higher than that at any other plant. Enteric virus levels in about 42% of the samples were below detection limits (0.3 to 8.3 MPN/100L), with some nondetects associated with all six facilities. Virus reductions were greatest in Facility B with the prechlorination (95%) and 66% of the samples were nondetects due to inactivation as opposed to physical removal. For Facilities C (lower loading with coagulation) and D (higher TSS in the effluent, higher loading rates and dual media), while virus reductions of 84% and 90% were achieved, 80% and 83% of the virus samples were positive. Minimal reductions were achieved at the other facilities.

A comparison of protozoan pathogen concentrations in the filtration effluents from the six wastewater reclamation facilities is shown in Figure 3-14. The concentration ranges for protozoan pathogens were similar for all facilities and ranged from about 1 to 10^3 cysts or oocysts/ 100 L. Mean *Giardia* concentrations were significantly different by facility, as levels at Facility A were significantly higher than any other facility. Of all the microbial parameters, only *Cryptosporidium* concentrations were not significantly different in filtered effluent by facility (P=0.3578). The concentration of *Giardia* was below detection limits (3-10.5 cysts/100 L) in about 13% of the filtered effluents, whereas the concentration of *Cryptosporidium* was below detection limits (1-11 oocysts/100L) in about 29% of the samples. All samples of filtered effluents from plants A, B, E, and F contained detectable levels of *Giardia*. All filtered effluent samples from plant A contained detectable levels of *Cryptosporidium*. Removals of *Giardia* cysts were 96-98.9% in facilities F, D and C. However, for the smaller oocysts even with coagulant added Facility C performed poorly (8% removal) and Facility F with the dual media depth filter with the greatest depth of all the filters performed the best (98.8% removal). The filters at the remaining facilities removed less than 60% of the oocysts.

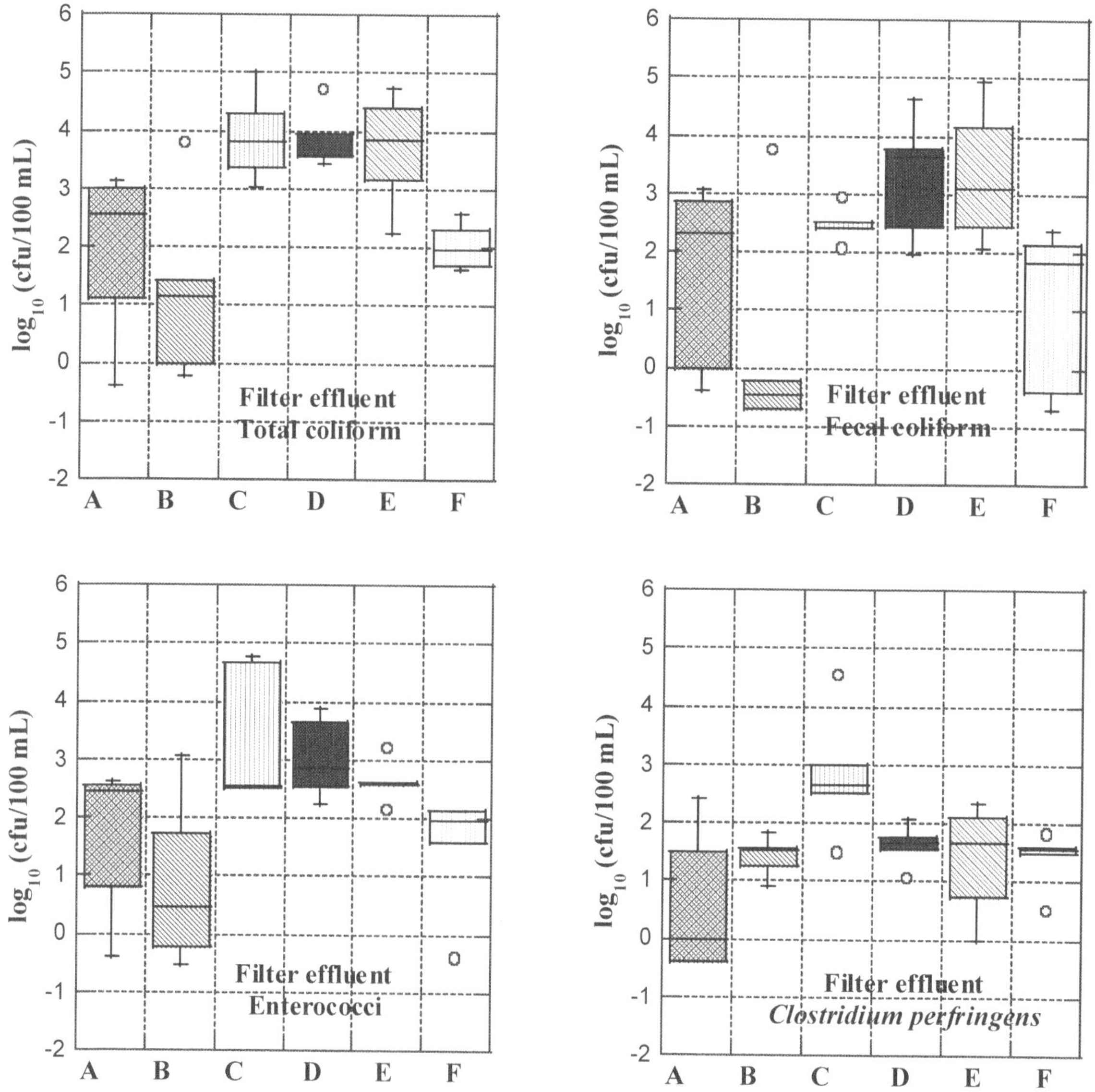

Figure 3-11. Comparison of Concentrations of Bacterial Indicators in Filtration Effluents from Each of Six Wastewater Reclamation Facilities. Facility Descriptions are Given in Table 2-1.

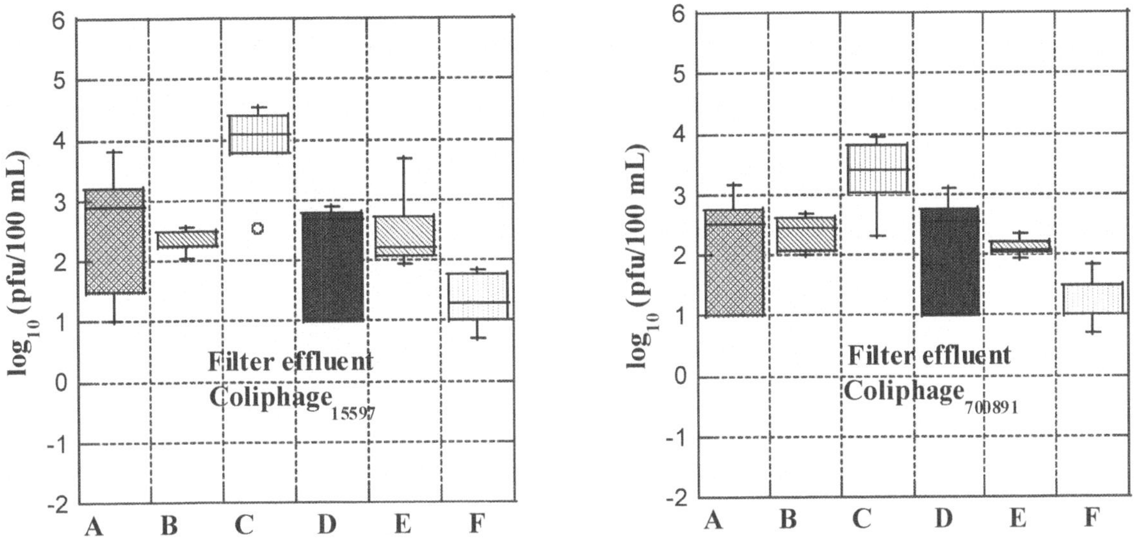

Figure 3-12. Comparison of Concentrations of Coliphages in Filtration Effluents from Each of Six Wastewater Reclamation Facilities. Facility Descriptions are Given in Table 2-1.

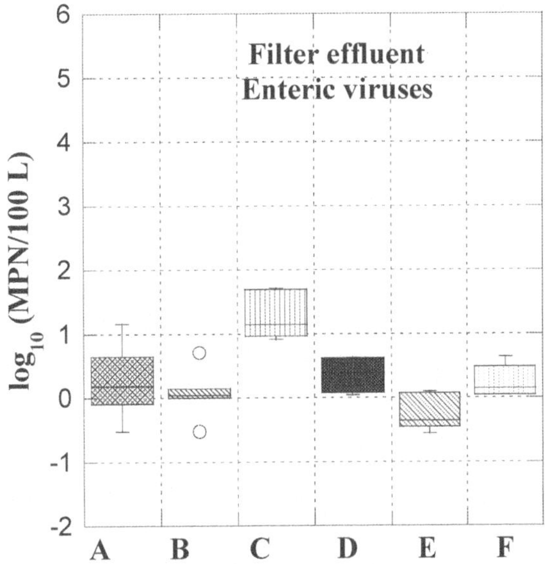

Figure 3-13. Comparison of Concentrations of Enteric Viruses in Filtration Effluents from Each of Six Wastewater Reclamation Facilities. Facility Descriptions are Given in Table 2-1.

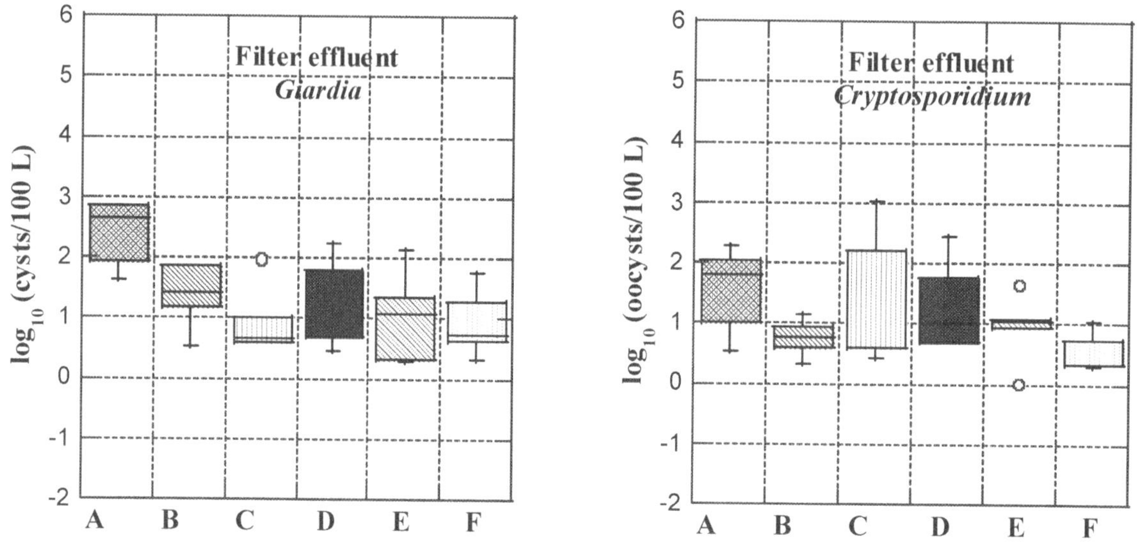

Figure 3-14. Comparison of Concentrations of Protozoan Pathogens in Filtration Effluents from Each of Six Wastewater Reclamation Facilities. Facility Descriptions are Given in Table 2-1.

3.3.2 Comparison of Process Variables

Filtration process variables that are significant include the filter media; use of upstream chemicals (oxidants and/or coagulants); hydraulic loading rates; operational variables such as filter run times and backwash practices; and water quality variables (turbidity, pH). One facility (B) uses prechlorination upstream of filtration, one facility uses a cationic polymer as a filter aide (C), and one facility uses a coagulant in the secondary sedimentation basin (F). The use of coagulant chemicals helps to increase the removal efficiency associated with filtration by modifying the surface characteristics and size of suspended particulates. If the indicators and pathogens are associated with other particulates, then removal through filtration is influenced by the size and surface characteristics of the bulk particulate matter. Specific data on turbidity and particle characteristics associated with each sample event were not available, thus it is not possible to analyze all of the potential process variables.

For the facilities tested in this project, four of the filters contained sand as either the only medium or in combination with anthracite in dual media filtration. The size of the sand varied among the facilities. The impact of the depth of sand in the filters on removal or indicators and pathogens was evaluated. A comparison of the relationship between sand depth and filter effluent concentrations is shown in Figure 3-15 for total coliform, enterococci, and *Cryptosporidium*. For these parameters, lower concentrations were associated with increased depth of sand in the filters.

The impact of filter depth on average detected concentrations and $\log_{10}$ reduction of total *Giardia* in the filter effluents from all facilities is shown in Figure 3-16. As shown, improved removal was associated with increased depth. Based on pooling the data from all facilities, there was an increase of about 0.014 $\log_{10}$ removal per inch of filter media. No trends were observed for other indicators or pathogens.

WERF

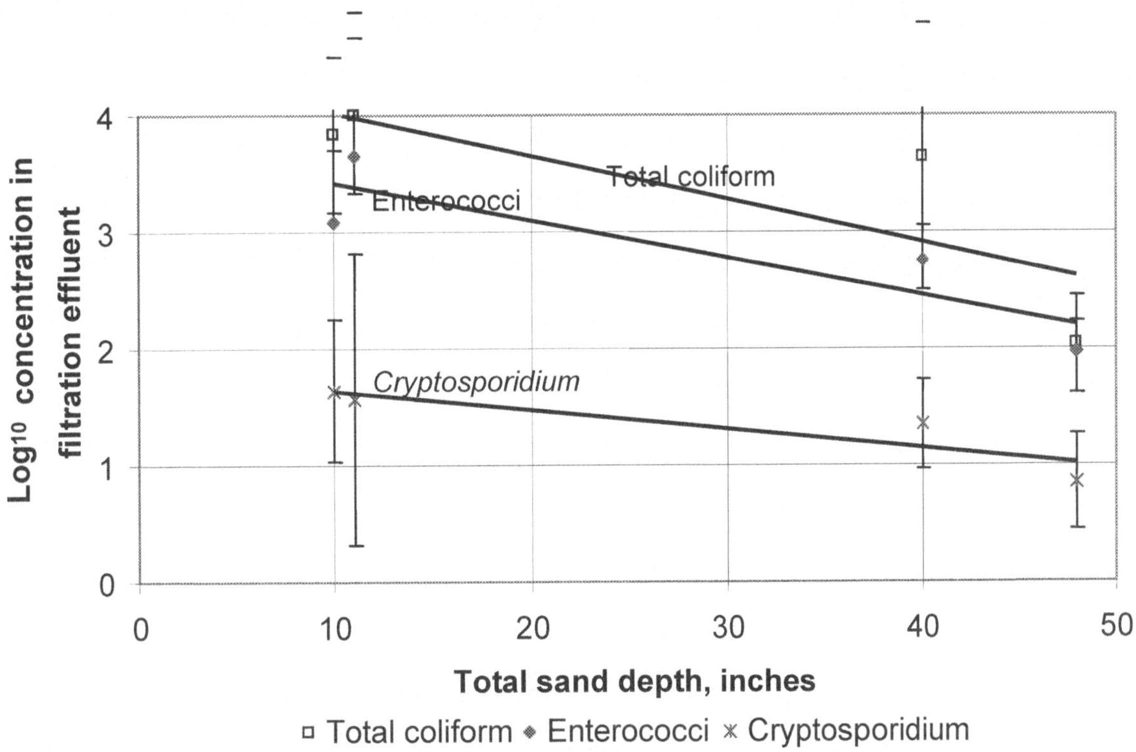

Figure 3-15. Impact of Sand Filter Depth on Llog$_{10}$ Average Detected Concentrations of Total Coliform (cfu/100 mL), Enteroocci (cfu/100 mL), and *Cryptosporidium*(oocysts/100 L) Through Secondary Effluent Filtration from Either Monomedia- or Dual-media Filtration with Sand. The correlation coefficients, r^2, are 0. 6 for total coliform and 0.8 for enterococci and *Cryptosporidium*.

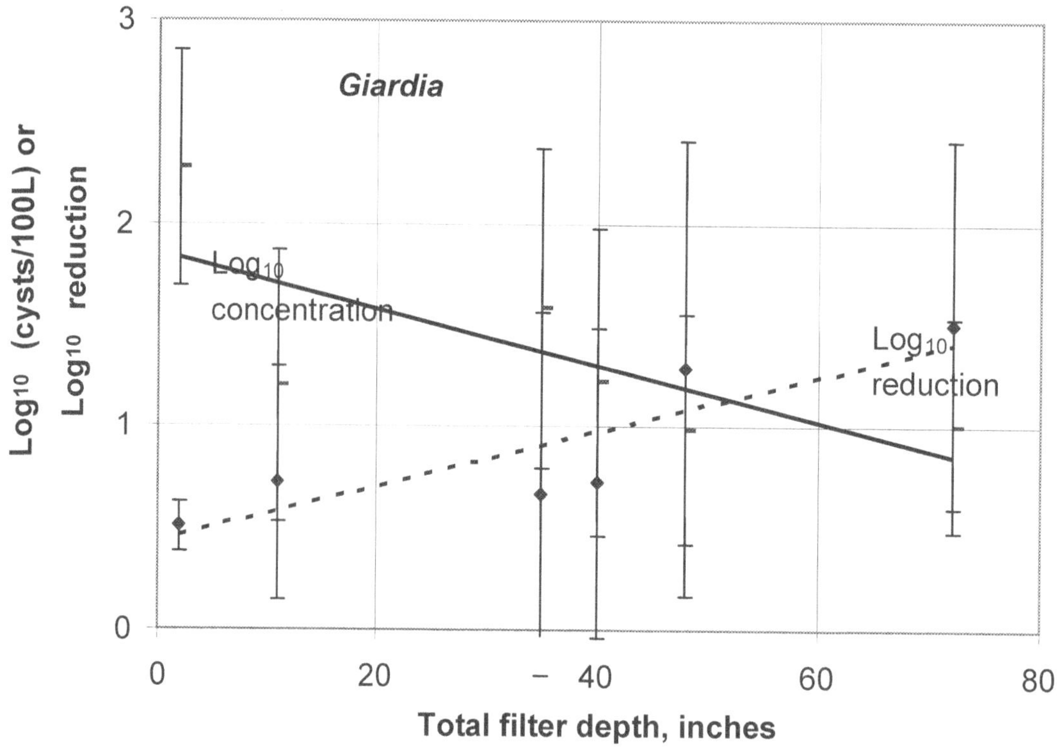

Figure 3-16. Impact of Nominal Filter Depth on Average Detected Concentrations and Log$_{10}$ Reduction of *Giardia* in Effluents from Cloth Ffilters, Monomedia- (sand or anthracite) and Dual-media Filters of Varying Depths. The correlation coefficient, r^2, is 0.5.

A comparison of the average log$_{10}$ reduction associated with filters that contain sand and the depth of sand is shown in Figure 3-17. There was an increase of about 0.03 log$_{10}$ reduction of *Cryptosporidium* and about 0.05 log$_{10}$ reduction of total coliforms per inch of sand. The graph does not include nondetected values. No correlations were observed for other indicators or pathogens tested in this project.

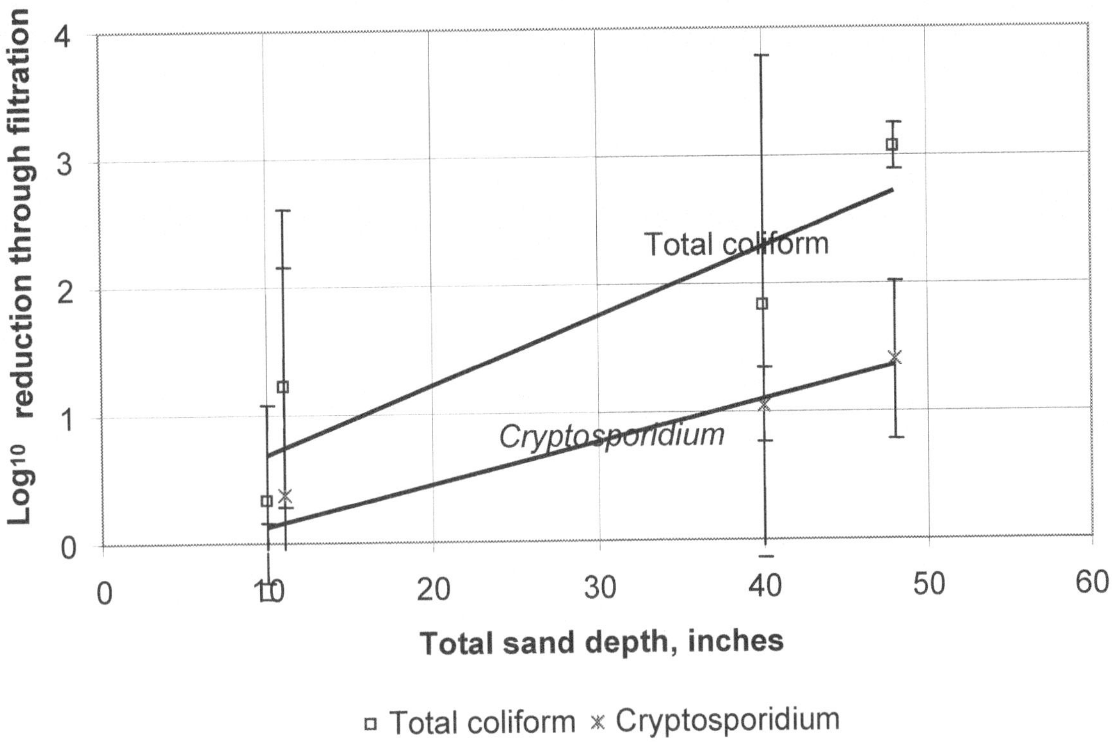

Figure 3-17. Impact of Sand Filter Depth on log10 Reduction of Average log Reduction of Total Coliform and *Cryptosporidium* through Secondary Effluent Riltration from either Monomedia or Dual Media Filtration with Sand. The Correlation Coefficients, r^2, are 0. 8 for Total Coliform and 0.9 for and *Cryptosporidium*.

The other filtration process variable that was evaluated in this project was hydraulic loading rate. Hydraulic loading rates varied from less than 1 for the automatic backwashing filter (Facility B) to over 5 gpm/ft^2 for the cloth filter (Facility A). A comparison of $\log_{10}$ reduction for all indicators and pathogens as a function of hydraulic loading rate is shown in Figure 3-18. Due to the wide range of operating conditions associated with the various filters, no strong correlations were observed for any of the individual microorganisms and hydraulic loading rates. Data are presented in Figure 3-19 for detected concentrations of total coliforms, fecal coliforms and enterococci in dual media filtration effluents. Higher concentrations of microorganisms were associated with higher hydraulic loading rates used in Facilities D and some extent E followed by F, with r^2=0.6 for total coliforms, 0.8 for fecal coliforms, and 0.6 for enterococci. No strong correlations were observed for other indicators and pathogens.

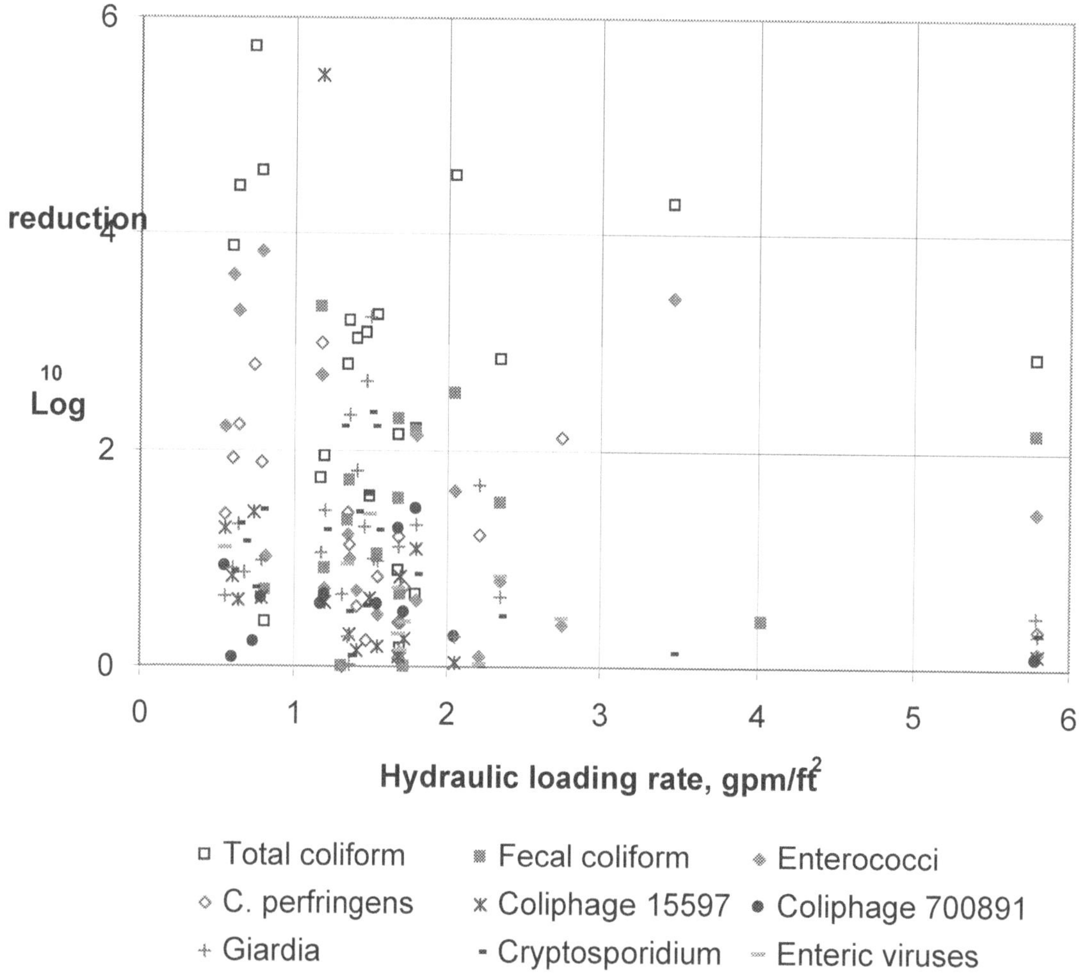

Figure 3-18. Comparison of $\log_{10}$ Reduction for Detected Concentrations of Indicators and Pathogens as a Function of Hydraulic Loading Rate for the Six Wastewater Reclamation Facilities Tested in this Project.

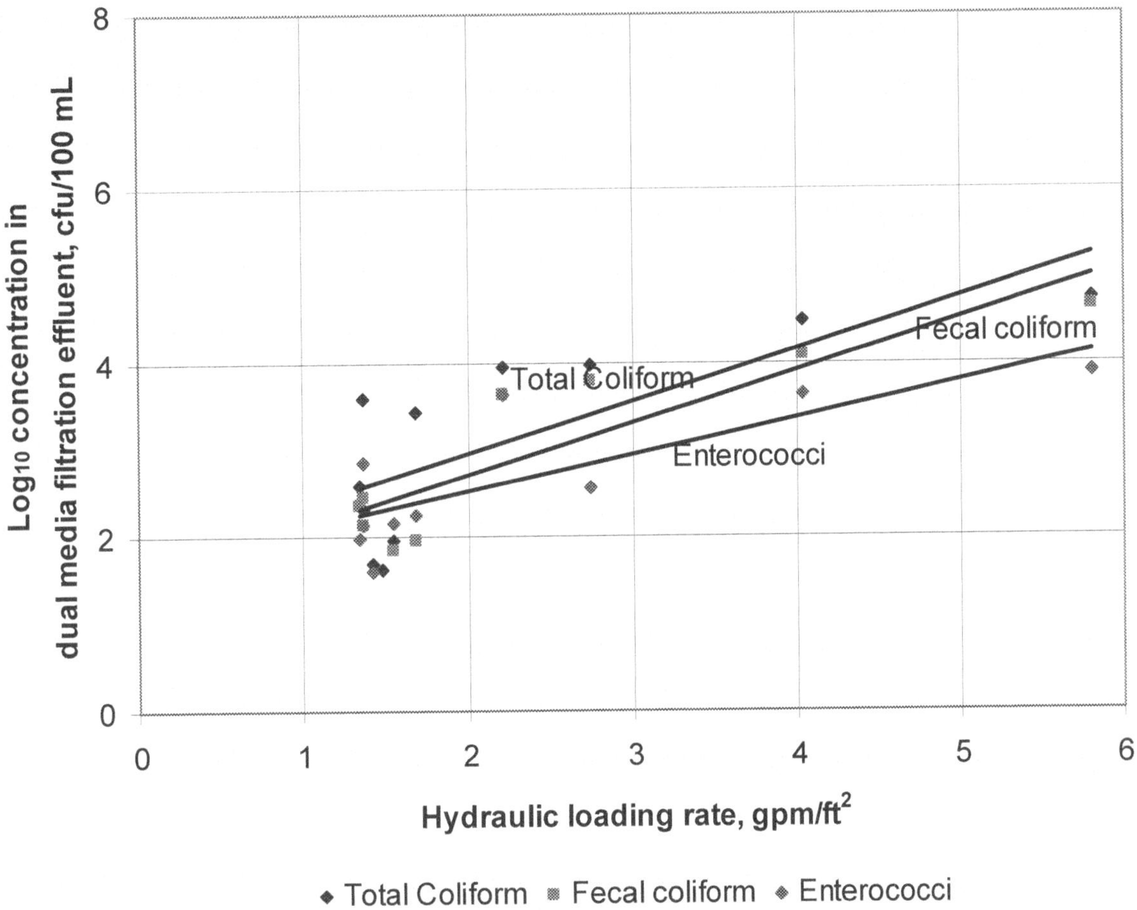

Figure 3-19. Comparison of Concentrations of Detected Concentrations of Bacterial Indicators (cfu/100 mL) in Dual-Media Filtration Effluents from Two Wastewater Reclamation Facilities (D & F) as a Function of Hydraulic Loading Rate. Correlation Coefficients, r^2, are 0.6 for Total Coliform, 0.8 for Fecal Coliform and 0.6 for Enterococci.

3.3.3 Impact of Filter Run Time and Prechlorination on Removal/Inactivation of Indicators

Intensive studies were conducted of the filtration process at Facilities B, C, and F. The intensive studies provided a means to compare the influence of filter media, run time, and upstream chlorination on removal of bacteria. A comparison of plant C (monomedia anthracite) and plant F (dual media anthracite and sand) is shown in Figure 3-20. As shown, the concentrations of indicator bacteria in the effluent from the anthracite media were consistently higher than the concentrations in the dual media effluent over the course of the filter runs and there were minimal variations in bacterial concentrations. Filter runs at Plant C are typically 48 hours and at Plant F are about 168 hours (see Table III-1). The data shown here represent the early stages of a filter run. The bacteria and coliphage levels were consistently higher in the secondary effluent in Facility C as compared to Facility F, and the removals by the filters averaged 99.65% compared to 97.7% for fecal coliforms, 76.9% compared to 90.7% for Enterococci and 99.2% compared to 46.5% for the somatic coliphage$_{15597}$ and 50% compared to 46% for the famp coliphage$_{700891}$. Facility C did utilize coagulation which may have aided in the greater removal of the somatic coliphage and the removals were greater with the anthracite for fecal coliform bacteria but was less with the round spherical enterococci.

In this case upstream processes (reduction of levels by secondary) and operations as well as design influenced the final water quality emerging from these two filters for different classes of indicators.

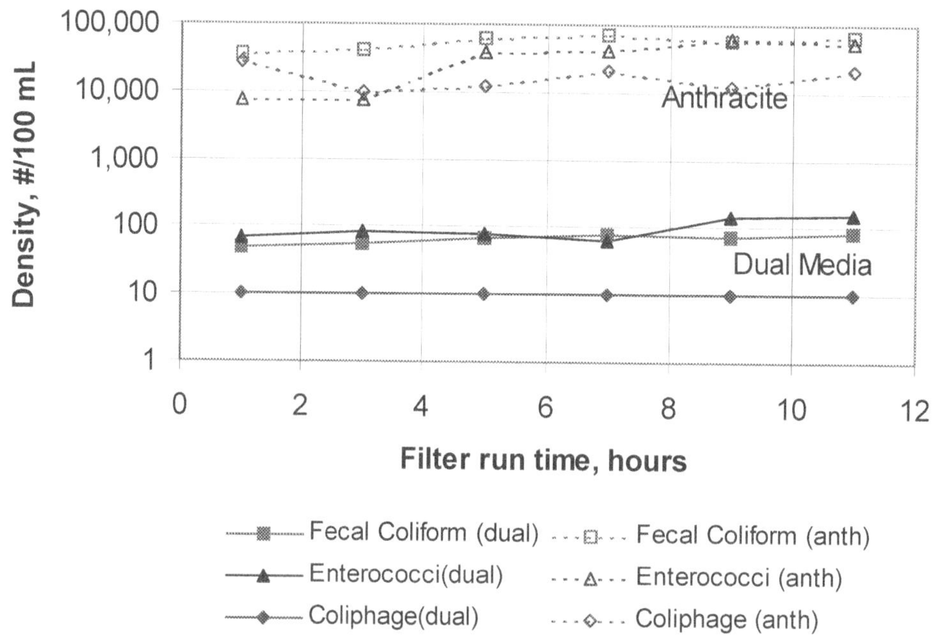

Figure 3-20. Comparison of the Concentration of Indicator Organisms in the Effluent from Anthracite Filters (Facility C) and Dual Media Filters (Facility F) During a Single Filter Run. Testing Conducted in Summer 2002.

A comparison of the effects of prechlorination on bacterial concentrations is shown in Figure 3-21. For fecal coliform and enterococci, there was a statistically significant difference in the concentration of bacteria with and without prechlorination. However, there was less of an impact on *C. perfringens*, perhaps because it is less susceptible to chlorine than the other bacteria assayed (Sobsey, 1989).

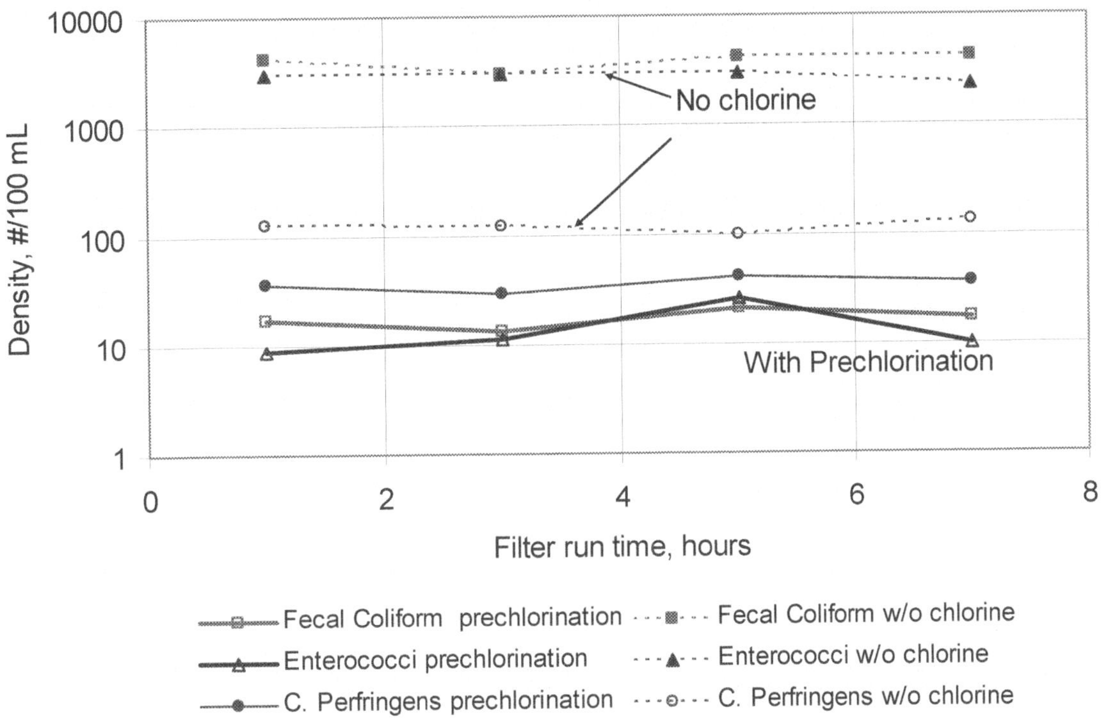

Figure 3-21. Comparison of the Concentration of Indicator Organisms in the Effluent from Shallow Sand Traveling Bridge Filters (Facility B) Operated With and Without Prechlorination. Testing Conducted in June 2002.

3.4 Reduction of Indicators and Pathogens Through Disinfection

The final treatment stage in the production of reclaimed water is disinfection. Five of the six facilities sampled during this project use chlorine disinfection and one (E) uses UV disinfection. Of the five facilities that use chlorine, one facility (F) practices nutrient removal and therefore has minimal ammonia available for reacting with the chlorine. The other four facilities (A, B, C, and D) do not routinely remove ammonia, thus the disinfection is most likely dominated by chloramines. The contact time available for disinfection varies among the treatment facilities and also with flow rates associated with each sampling event.

3.4.1 Concentrations of Indicators and Pathogens in Disinfected Effluents

A comparison of the concentrations of bacteria in disinfected effluents is shown in Figure 3-22. Median concentrations of total coliforms ranged from about 0.3 to 10 cfu/100 mL, while fecal coliforms, enterococci, and *Clostridium perfringens* averaged about 1 cfu/100 mL. The highest concentrations of total coliforms were associated with disinfected effluents from Facility E (UV disinfection) and the highest average concentrations of fecal coliforms and enterococci were associated with Facilities D and E. The highest concentrations of *Clostridium perfringens* were associated with Facility C. Keeping in mind that larger volumes than the routine 100 ml were tested, the Total coliform levels were below detection limits (0.3 to 1 cfu/100 mL) in about 37% of the disinfected effluent samples. Fecal coliforms and enterococci were below detection limits (0.3 to 1 cfu/100 mL) in about 73% of the samples, and *Clostridium perfringens* was below detection limits in about 39% of the samples.

In meeting the standard of <1 fecal coliform bacteria/ 100 ml, facilities A, B, C and F would have done so 100% of the time and Facilities D and E would have done so only 66% and 33% of the time. Only Facility C would have met the standard of <1 total coliform bacteria /100ml in 75% of the samples collected. The other facilities had greater than that level in 40-80% of the samples collected.

A comparison of concentrations of coliphages in disinfected effluents is shown in Figure 3-23. Coliphages were below detection limits (5-10 pfu/100 mL) in about 55-62% of the samples. Detected concentrations ranged from 1 to 100 pfu/100 mL with the highest concentrations associated with Facility C.

Concentrations of enteric viruses in disinfected effluents are compared in Figure 3-24. Enteric viruses were below detection limits (0.3 to 1.5 MPN/100L) in 69% of the samples. They were not detected in this study in the disinfected effluent from Facility E (UV) and Facility B (prechlorination of filter) rarely in D. In most cases, the detected concentrations of enteric viruses were below 1 MPN/100L.

A comparison of the concentrations of total *Giardia* and total *Cryptosporidium* are shown in Figure 3-25. The concentration of *Giardia* was highest in effluents from plant A and similar for the other facilities with median levels around 1 cyst/100 L. The concentration of *Cryptosporidium* was highest in facilities C and D. *Giardia* concentrations were below detection limits (1.8-5.2 cysts/100 L) in 17% of the disinfected effluent samples. *Cryptosopridium* concentrations were below detection limits (2-6.9 oocysts/100 L) in 30% of the disinfected effluent samples.

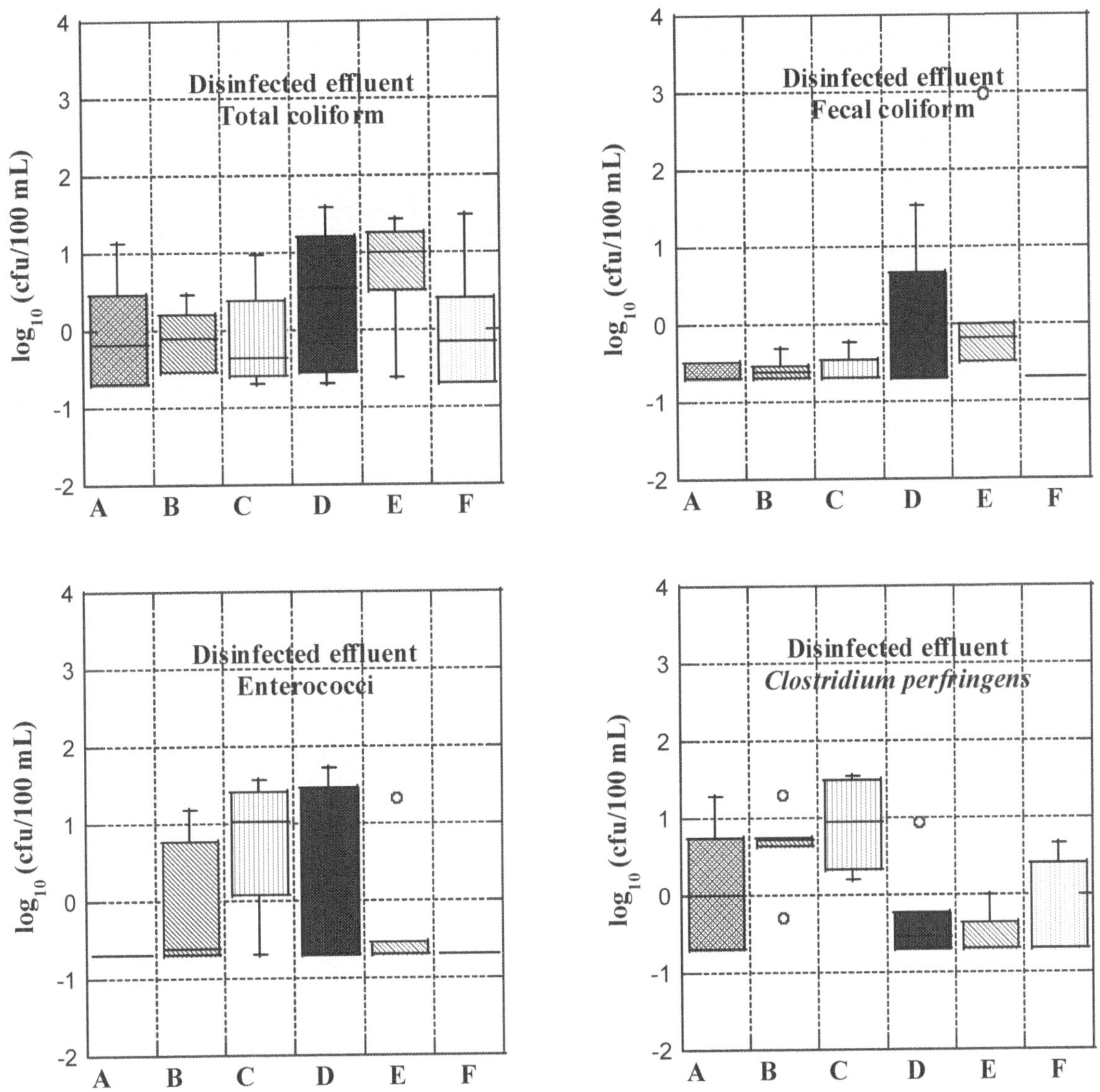

Figure 3-22. Comparison of Concentrations of Bacterial Indicators in Disinfected Effluents from Each of Six Wastewater Reclamation Facilities. Facility Descriptions are Given in Table 2-1.

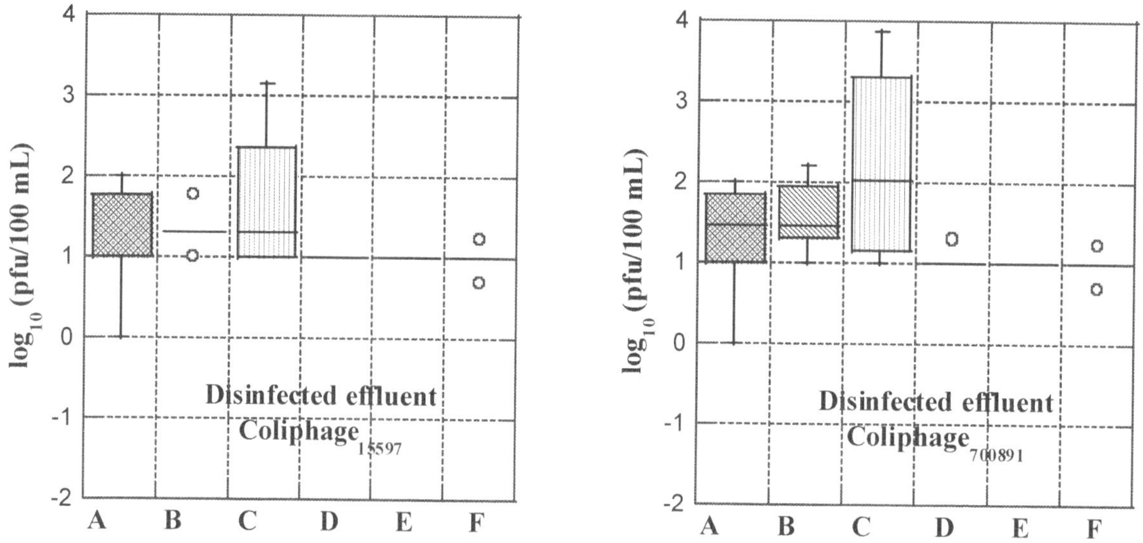

Figure 3-23. Comparison of Concentrations of Coliphages in Disinfected Effluents from Each of Six Wastewater Reclamation Facilities. Facility Descriptions are Given in Table 2-1.

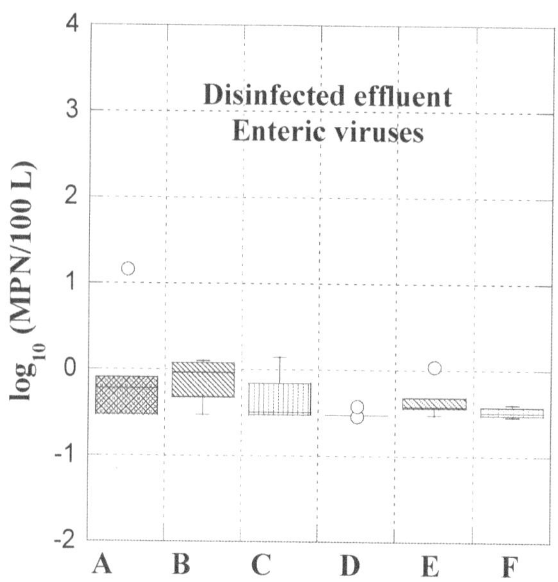

Figure 3-24. Comparison of Concentrations of Enteric Viruses in Disinfected Effluents from Each of Six Wastewater Reclamation Facilities. Facility Descriptions are Given in Table 2-1.

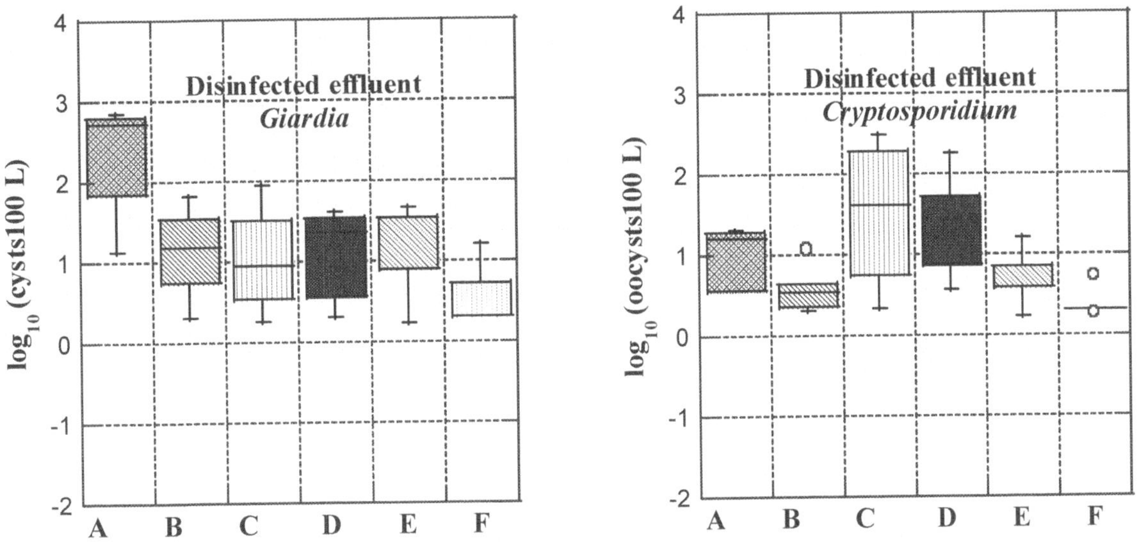

Figure 3-25. Comparison of the Concentrations of Total Counts of *Giardia* (cysts/100L) and *Cryptosporidium* (oocysts/100L) in Disinfected Effluents from Each of Six Wastewater Reclamation Facilities. Facility Descriptions are Given in Table 2-1.

3.4.2 Impact of Process Variables on Removal of Indicators and Pathogens Through Disinfection

The effectiveness of disinfection is impacted by a number of process variables including disinfection dose, contact time, temperature, and water quality variables (pH, turbidity, presence of ammonia, oxidant demand). Due to the design of this study and data limitations, not all process variables could be evaluated. A comparison of the impact of contact time and $\log_{10}$ reduction of enterococci and enteric viruses is shown in Figure 3-26. As would be expected, increased removal is associated with longer chlorine contact times. The slopes for the two lines are similar with an increase of about 0.006 $\log_{10}$ reduction per additional minute of contact time.

A comparison of the impact of secondary effluent ammonia levels on the detectable concentrations of enterococci and fecal coliforms is shown in Figure 3-27. For these indicators, higher levels of ammonia were associated with higher enterococci concentrations in the disinfected effluents.

One of the facilities tested during this study uses UV disinfection (Facility E in Table 2-1). The UV disinfection system (Trojan UV4000) consists of two UV reactors in series. Each reactor contains two banks of medium pressure, mercury vapor hot cathode-instant start lamps with eight modules per bank and eight lamps per module (256 lamps per UV channel). Each channel can treat an average flow of 8 MGD and a peak flow of 14.4 MGD. The reactors have a horizontal lamp configuration and the flow is parallel to the lamps. The lamps are enclosed in cylindrical Type 214 quartz sleeves which allow 89% UV transmittance. The intensity of the medium pressure lamps is adjusted based on UV transmittance and turbidity. A comparison of filtered turbidity and $\log_{10}$ reduction of total coliform, fecal coliform and *Giardia* is shown in Figure 3-28. As shown, higher levels of turbidity are associated with decreased removal of the indicators. For the total and fecal coliform, the $\log_{10}$ reduction decreased about 0.5 per unit of turbidity (NTU). The *Giardia* numbers represent total counts and do not reflect infectivity but just like turbidity are an indicator of the effectiveness of the filter process prior to disinfection.

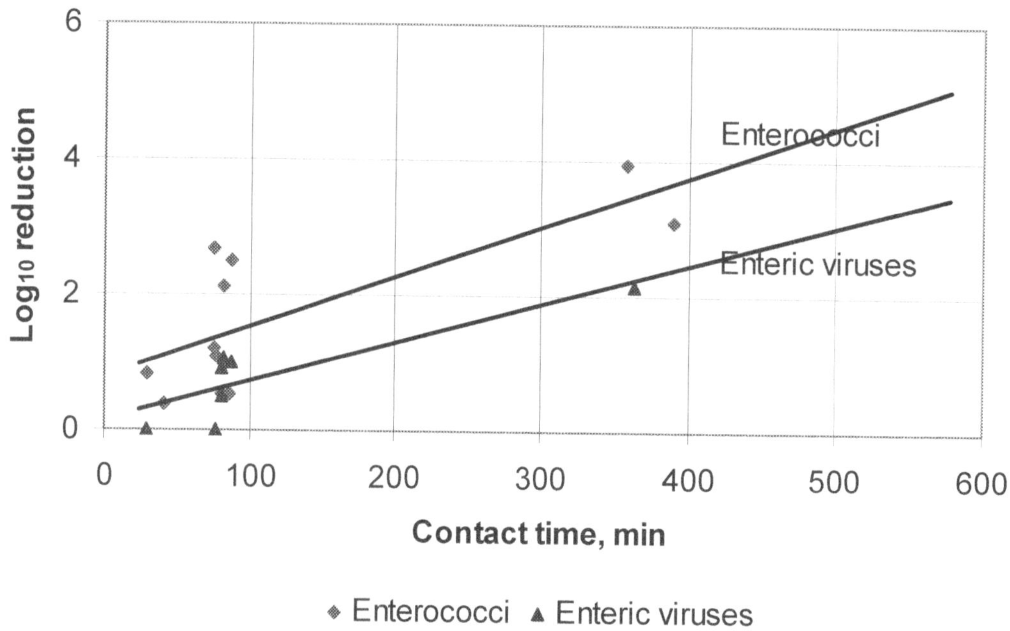

Figure 3-26. Impact of Chlorine Contact Time on $\log_{10}$ Reduction of Enterococci (r^2=0.6) and Enteric Viruses (r^2=0.7). Data Plotted Represent Samples that had Detectable Levels of Each Parameter.

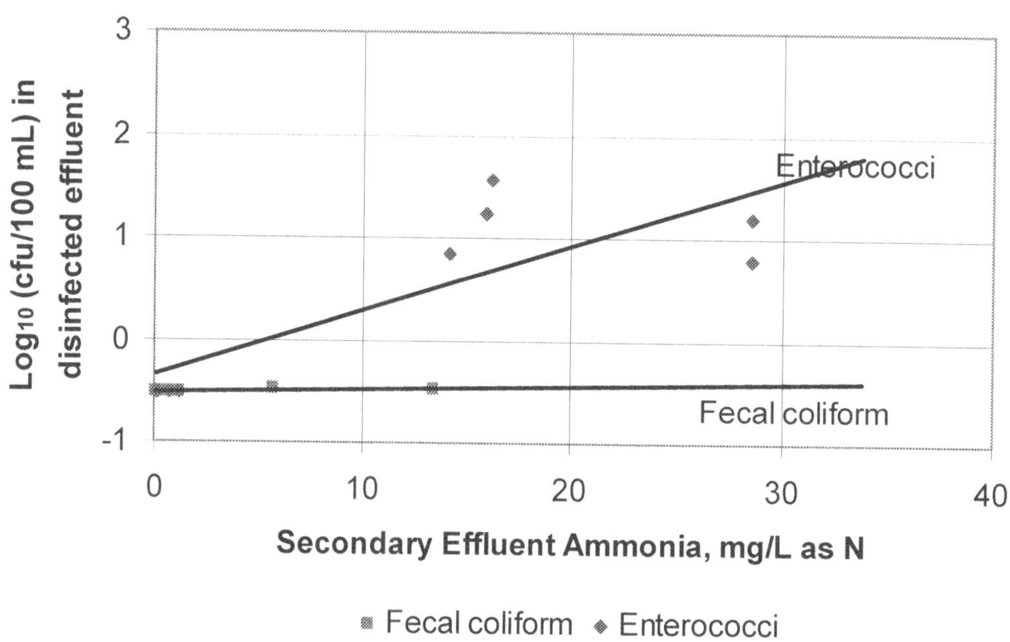

Figure 3-27. Comparison of Impact of Secondary Effluent Ammonia on Detectable Concentrations of Enterococci (r^2=0.7) and Fecal Coliforms(r^2=0.8) in Disinfected Effluents.

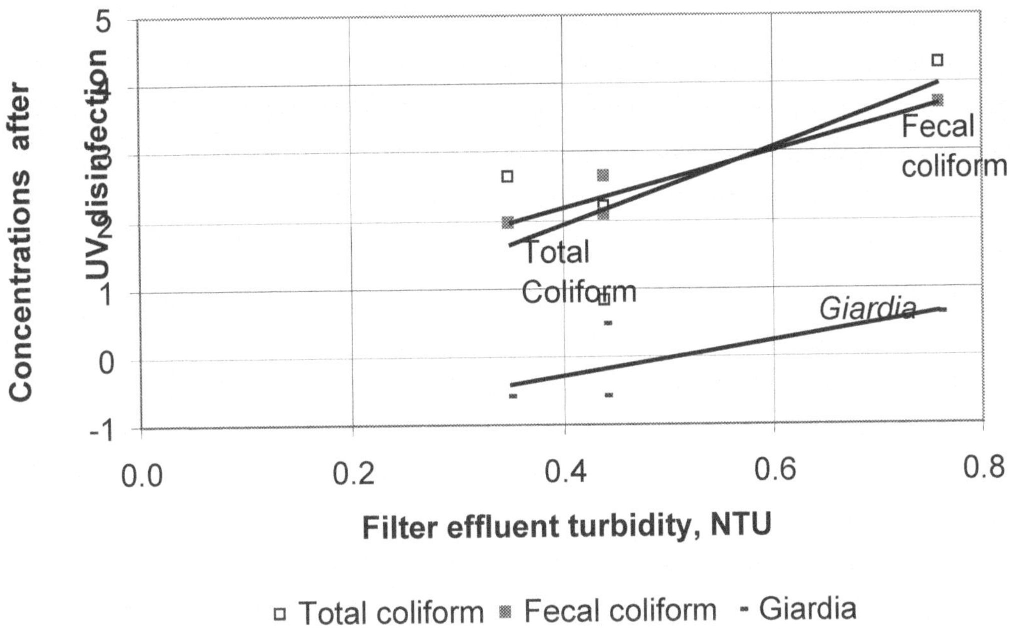

Figure 3-28. Comparison of Filtered Effluent Turbidity on Concentrations of Total Coliform, Fecal Coliform and *Giardia* Through UV Disinfection. Correlation Coefficients, r^2, are 0.5 for Total Coliform and *Giardia* and 0.9 for Fecal Coliform.

3.5 Protozoan Pathogen Viability

Specialized tests were conducted to evaluate the patterns of viability associated with monitoring of protozoan pathogens. *Giardia* viability was tested using fluorogenic dyes. *Cryptosporidium* viability or infectivity was tested using an additional culturing step.

3.5.1 *Giardia* Cyst Viability

The inclusion or exclusion of the fluorogenic dyes 4'-6-diamidino-2-phenylindole (DAPI) and propidium iodide (PI) has been used as a marker of intact membranes in waterborne *Cryptosporidium* and *Giardia* and as an indicator of the presence of internal features such as nuclei (Schupp and Erlandsen, 1987; Campbell et al., 1992). PI is capable of passing through only damaged cell membranes, and intercalates with the nucleic acids of injured and dead cells to form a bright red fluorescence complex (Sauch et al., 1991). DAPI is an AT-selective DNA stain, which causes a 20-fold enhancement in fluorescence when binding to DNA occurs.

The fluorogenic vital dye assay or dye permeability assay tests the differential uptake of the fluorochromes 4'-6-diamidino-2-phenylindole (DAPI) and propidium iodide (PI) and thus those cysts that take in PI are considered nonviable.

In this project, the combined use of PI and the fluorescein labeled detection method was applied for identifying *Giardia* cysts. When DAPI was used alone and with FITC labeled antibodies, two population of cysts were observed: (i) cysts with two or four blue dots (DAPI+) when observed under fluorescent microscopy with the UV filter block, suggesting that intact nuclei of the trophozoites were stained with the fluorochrome. The same population presented internal morphological characteristics typical of *Giardia* when viewed under DIC microscopy (peritrophic space, axonema); (ii) cysts presenting an apple green fluorescence when viewed with the blue filter block; these cysts did not show any internal morphological characteristics under DIC microscopy. No spikes, stalks, appendages, pores, large nuclei filling the cell, red fluorescing chloroplasts, crystals or spores were observed, therefore these apple-green fluorescing objects were considered as empty *Giardia* cysts.

When PI was used alone and with FITC labeled monoclonal antibodies, two populations were observed: i) cysts which included PI staining (PI+) and looked bright red with two or four visible internal nuclei when examined with the green filter block and with an external apple green fluorescence along with a reddish color diffused into the cyst when examined with the blue filter block ii) cysts with an external apple green fluorescence when examined with the blue filter block but empty and with no internal morphological characteristics when examined under DIC.

In tests in which DAPI and PI were used with FITC, four populations were observed: i) the same type of cysts stained bright red with two or four visible internal nuclei when examined with the green filter block (PI+) and with an external apple green fluorescence along with a reddish color when examined with the blue filter block, ii) empty cysts (DAPI-, PI-); (iii) very low numbers of cysts including PI into their nuclei only, (iv) a very reduced number of cysts showing a diffuse DAPI staining that were PI-.

It appears that both DAPI and PI provide an additional confirmatory step to determine the presence of *Giardia* cysts in wastewater samples and to distinguish the protozoan parasite from other nontarget organisms that can be recovered during sample collection and concentration/purification procedures. Some researchers have reported that the exclusion of the fluorogenic dye PI provides a general determination of living cells with intact cell membranes (Shupp and Erlandsen, 1987). In this study, none of the cysts that excluded either PI alone or the fluorochromes DAPI/PI showed internal morphological characteristics of *Giardia* cysts when viewed under DIC, therefore these cysts were considered as empty cysts. Jenkins et al., (1997) consider that the dye permeability assay provides an economical method to determine viability and potential infectivity, although the estimate may be conservative. The effect of the acid treatment during immunomagnetic separation was not tested in this project.

In the second year of the study PI and DAPI were incorporated into the protocol for IFA. The percentage of *Giardia* cysts that were DAPI+/PI- indicating viability in the total population of cysts were as follows: For Facility A, 9.1%, 1.5%, 4.7%, and 2.4% of the cysts were viable in untreated sewage, secondary effluent, filtered effluent and disinfected final reclaimed effluent, respectively. For Facility D, 4.3%, and 3.2% of the cysts were viable in untreated sewage and final disinfected final reclaimed effluent, respectively [none were detected in secondary and filtered] and for Facility E, 1.2% and 2.8% of the cysts were viable in untreated sewage and secondary effluent, respectively[no cysts detected in filtered effluent and disinfected final reclaimed effluent].

Viable *Giardia* Cysts were detected in two of the facilities' disinfected effluent (A and D) and the data indicate that little reduction of "viability" is achieved through treatment. It should be kept in mind that the DAPI/PI can not be used to indicate viability after UV disinfection.

A comparison of the concentration of detected DAPI/PI- *Giardia* cysts and total detected *Giardia* cysts is shown in Figure 3-29 (n=8). While the number of positive samples is low, this relationship suggests that removal of *Giardia* cysts through the treatment steps tested in this project (biological treatment, filtration, and disinfection) is not impacted by cell viability.

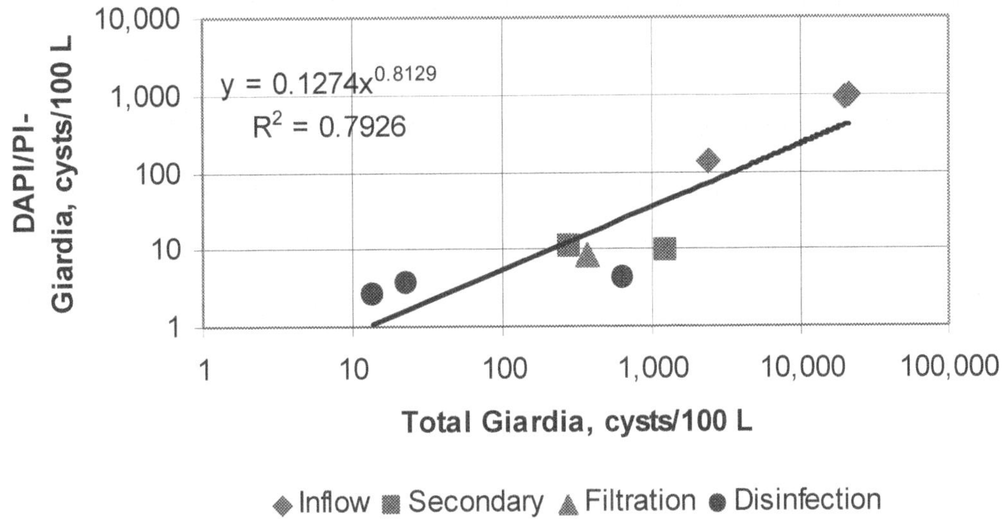

Figure 3-29. Comparison of the Concentration of Detected DAPI/PI- Giardia Cysts and Detected Concentrations of Total Giardia Cysts from Different Stages of Treatment and DifferentTreatment Facilities. The Correlation Coefficient for the Relationship is 0.8.

3.5.2 *Crytosporidium* Oocyst Viability

Use of U.S. EPA Method 1623 for detection of *Cryptosporidium* oocysts does not permit determination of oocyst viability or infectivity. Testing for the presence of viable or infectious oocysts in the final effluent is important for uses of reclaimed water where ingestion is expected to occur. If the oocysts are not viable or infectious, health risks from exposure to *Cryptosporidium* oocysts are minimal. However, if the disinfection process fails in inactivating oocysts, viable oocysts may be released to the environment in the reclaimed water. In this study the use of cell culture methods were used to determine the concentrations of infectious oocysts.

In facilities A, B, C, D, E, and F, infectious oocysts were detected in 20%, 40 %, 0%, 33%, 0%, and 50% of the final reclaimed water effluent samples, respectively. Levels ranged from 2.5-8 oocysts/100L. While it is suspected based on the literature that uv disinection employed in Facility E would inactivate the oocysts, many more samples would need to be collected and assayed to prove this statement in a full-scale facility.

A comparison of the number of infectious *Cryptosporidium* oocysts (MPN•100 L^{-1}) and total *Cryptosporidium* oocysts (oocysts •100 L^{-1}) is shown in Figure 3-30 for data sets in which both measurements were above the detection limits (n=15). The proportion of infectious oocysts is relatively consistent in the influent samples, but the relationship is more difficult to distinguish after the filtration treatment step due to low limits of detection and limited positive samples. For the pooled data from all plants, the empirical relationship for the infectious oocysts in the influent and secondary effluent can be modeled as (r^2=0.8):

$$\text{Infectious oocysts} = 1.46*(\text{total oocysts})^{0.8}$$

There was no observed correlation between infectious and total oocysts in the filtered effluents and disinfected samples (n=7) (r^2=0.006). In three of the samples about 100% of the oocysts detected by IFA were found to be infectious and four other sample comparisons showed that less than 10% of the total oocysts were infectious.

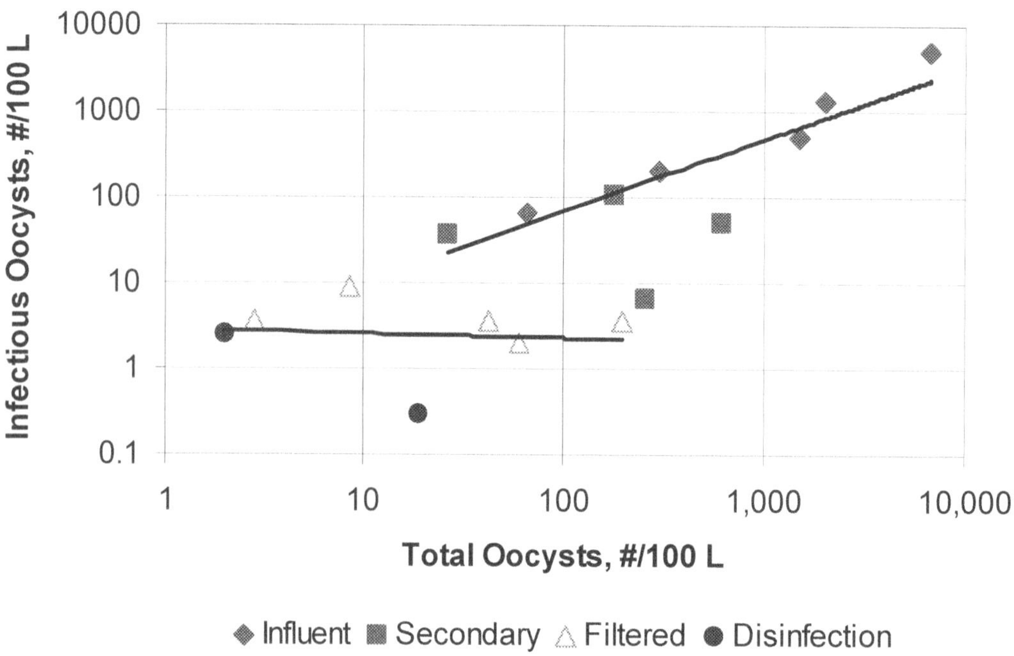

Figure 3-30. Comparison of Detected Levels of Infectious Oocysts and Total Oocysts from Samples of Untreated, Secondary Effluent, Filtered Effluents, and Disinfected Effluents from Six Wastewater Reclamation Facilities.

CHAPTER 4.0

DISCUSSION: ANALYSIS OF INDICATORS AND PATHOGENS

Typically, indicator organisms total and fecal coliform bacteria concentrations are the only monitoring tool used to assess the microbial safety of reclaimed water. Utilities and regulatory agencies rely upon the assumed predictive relationship between indicator organism and pathogen survival/transport through wastewater treatment to ensure that the reclaimed water is safe for public use. While it is impossible to test reclaimed waters for all possible pathogens, it is important that the indicator organism(s) used to ensure water quality should be correlated with a broad variety of waterborne pathogens, i.e. bacteria, viruses and protozoa.

Six wastewater treatment plants that produce reclaimed water were each sampled at least four times over the course of this project. The levels of indicator organisms detected in the reclaimed water produced by each of the facilities are all within the permit requirements for each facility (total or fecal coliforms). Because sampling for this project was limited to peak flow conditions, the data provide a conservative perspective on the capacity of each facility for pathogen reduction.

4.1 Patterns of Detection and Non-detection

A comparison of the percent of samples that contained detectable levels of indicators is shown in Figure 4-1. Over 80% of all influent, secondary, and filtration samples contained detectable levels of total and fecal coliforms and enterocci. Following disinfection, the most persistent indicators were total coliform and *Clostridium perfringens* which were present in over 60% of the samples. Fecal coliforms were below detection levels in all disinfected effluent samples from plants B and C, and both coliphage indicators were below detection levels in the final effluent from the facility that uses UV disinfection (E). Total coliforms were above detection limits for all final effluent samples from the two facilities that have nitrogen removal (E and F), whereas *Clostridium perfringens* was above detection limits in all samples from the facility with the longest chlorine contact time (C). It should be noted that larger sample volumes for coliforms in effluent samples were used in this study compared to routine monitoring that is conducted for wastewater reclamation facilities, resulting in lower detection limits. Lower limits of detection were used for coliphage (10ml samples volumes). Fecal coliforms and enterococci were detected in less than half of the samples and were apparently most sensitive to treatment of all the indicators evaluated.

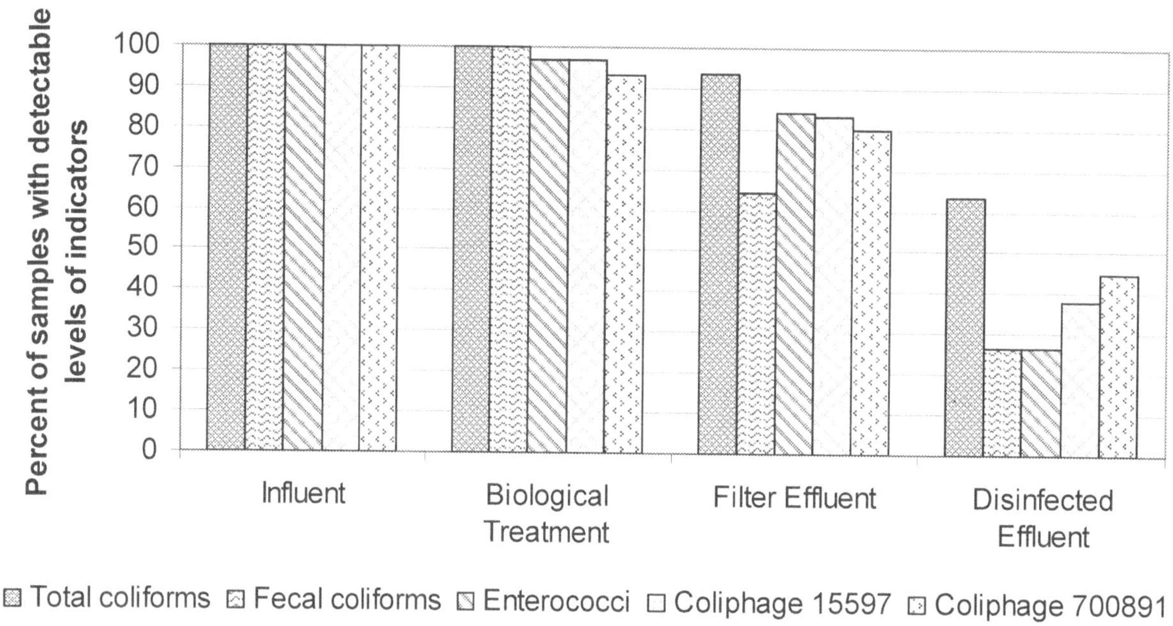

Figure 4-1. Comparison of the Influence of Treatment on Detection of Indicators from Untreated and Treated Wastewater at Six Wastewater Reclamation Facilities.

A comparison of the patterns of detection associated with pathogens tested in this project is shown in Figure 4-2. As shown, all untreated wastewater samples contained detectable levels of enteric viruses and *Giardia*, whereas about 30% of the final effluent samples contained detectable levels of enteric viruses. All samples of disinfected effluents from plants B (prechlorination of filter) and E (UV disinfection) did not contain detectable levels of enteric viruses. The limits of detection associated with quantification of *Giardia* and *Cryptosproridium* varied with sample location, therefore patterns of detection/nondetection tended to be less consistent and did not correspond to removal patterns for any of the indicators. *Giardia* were detected in over 80% of the disinfected effluent samples, with the highest frequency of non-dectection associated with the biological nutrient removal facility (plant F). Viable *Cryptosproridium* were not detected in any of the disinfected effluents from plant A (cloth filter), C (longest chlorine contact times), or E (UV disinfection).

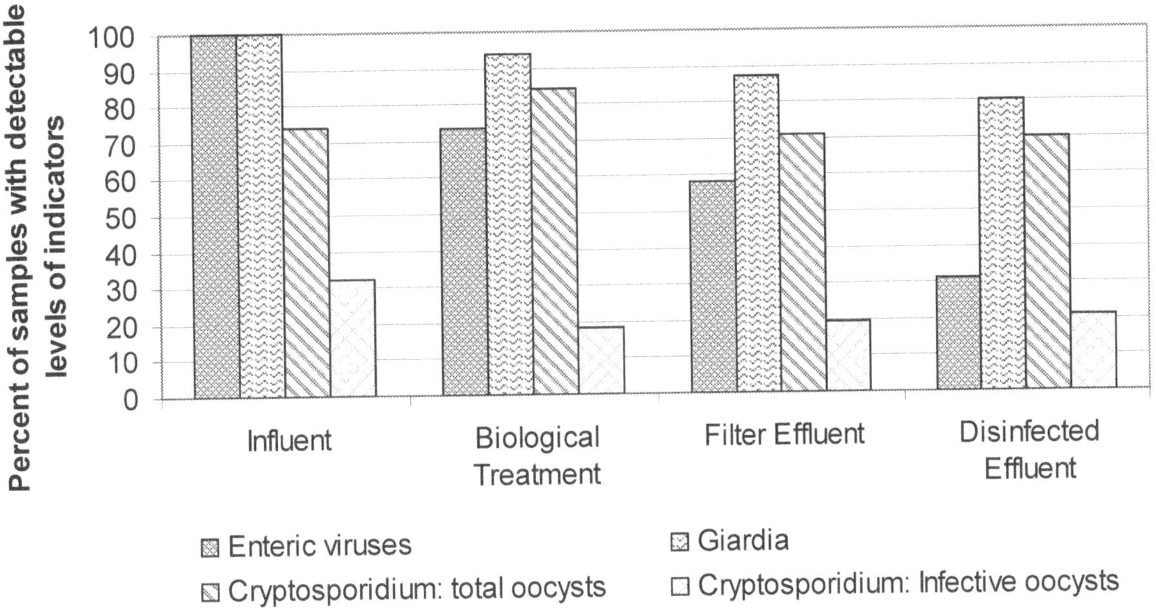

Figure 4-2. Comparison of the Influence of Treatment on Detection of Indicators from Untreated and Treated Wastewater at Six Wastewater Reclamation Facilities.

No direct correlations were observed between combination of indicator(s) and pathogen(s) for the data collected through this study, however, correlations were observed among the indicators. A comparison of detected levels of total and fecal coliform is shown in Figure 4-4. While total coliforms tend to be more persistent than fecal coliforms, when fecal coliforms are above detection limits, there is a consistent relationship through each treatment step.

A comparison of the detected levels of fecal coliform and enterococci is shown in Figure 4-3. Fecal coliforms were below detection limits in all disinfected effluent samples from plants B and C, while enterococci were below detection limits in all disinfected effluent samples from plant A. Both indicators were only codetected in one sample from plant D and four samples from plant F. When both indicators were codetected there was a consistent relationship between their detected concentrations.

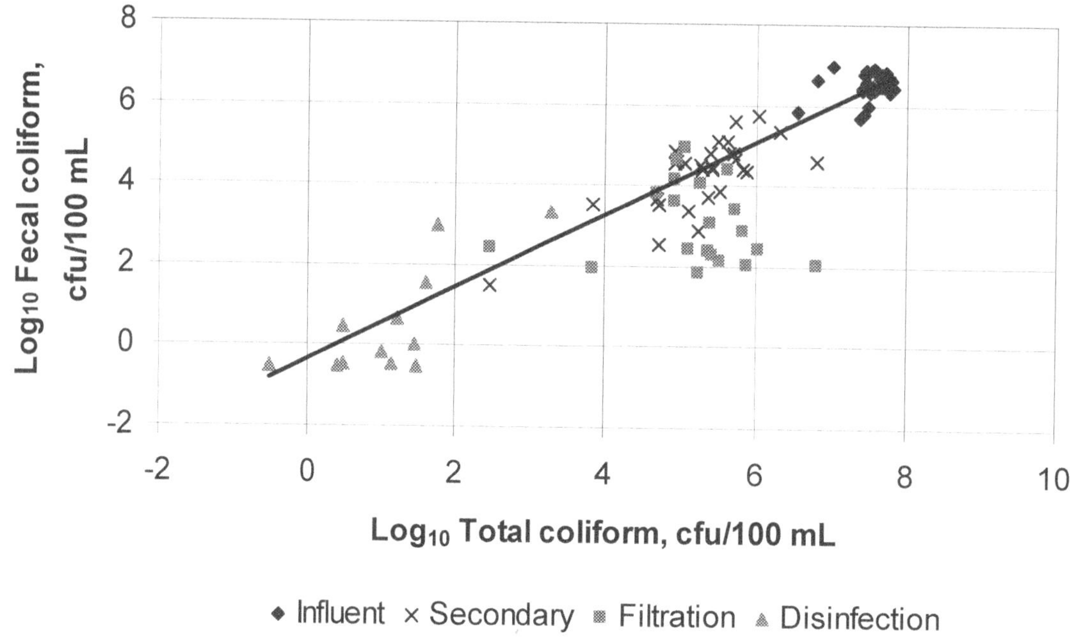

Figure 4-3. Comparison of Detected Levels of Total and Fecal Coliforms in Samples of Untreated Wastewater, Secondary Effluent, Filtration Effluents, and Disinfected Effluents from Six Wastewater Reclamation Racilities. The Correlation Coefficient, r^2, is 0.9 and the Slope of the Line is 0.9.

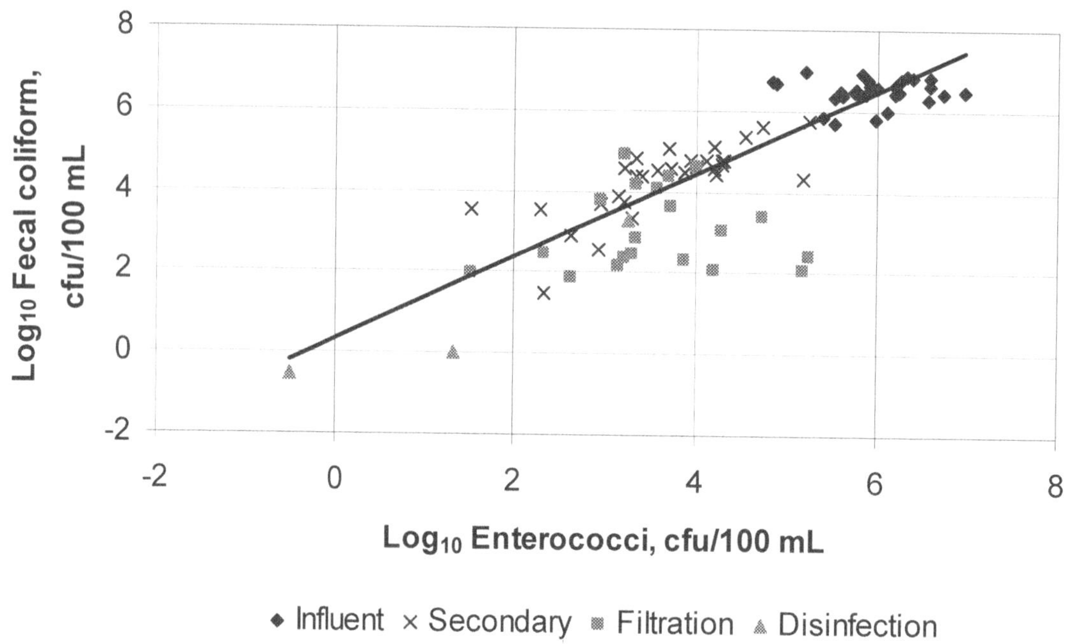

Figure 4-4. Comparison of Detected Levels of Enterococci and Fecal Coliforms in Samples of Untreated Wastewater, Secondary Effluent, Filtration Effluents, and Disinfected Effluents from Six Wastewater Reclamation Facilities. The Correlation Coefficient, r^2, is 0.9 and the Slope of the Line is 1.

Bacteriophages have been suggested as alternative indicator organisms for enteric viruses as their morphology and survival characteristics resemble some of the enteric viruses (Turner and Lewis, 1995). Havelaar et al., (1993) showed that for monitoring purposes F-specific coliphages can serve as surrogates of human pathogenic enteric viruses in ambient waters. In this study a comparison of detected levels of coliphages and enteric viruses in samples of disinfected effluents with a variety of effluent qualities from all six facilities tested is shown in Figure 4-5. While no correlation between the numbers of coliphages and enteric viruses was found (coliphages can not be used to correlate to the levels of enteric viruses), it may be that one can predict the absence of enteric viruses and based on the data, as levels less than 10 coliphage/100ml were indicative of effluents with no detectable cultivatable enteric viruses.

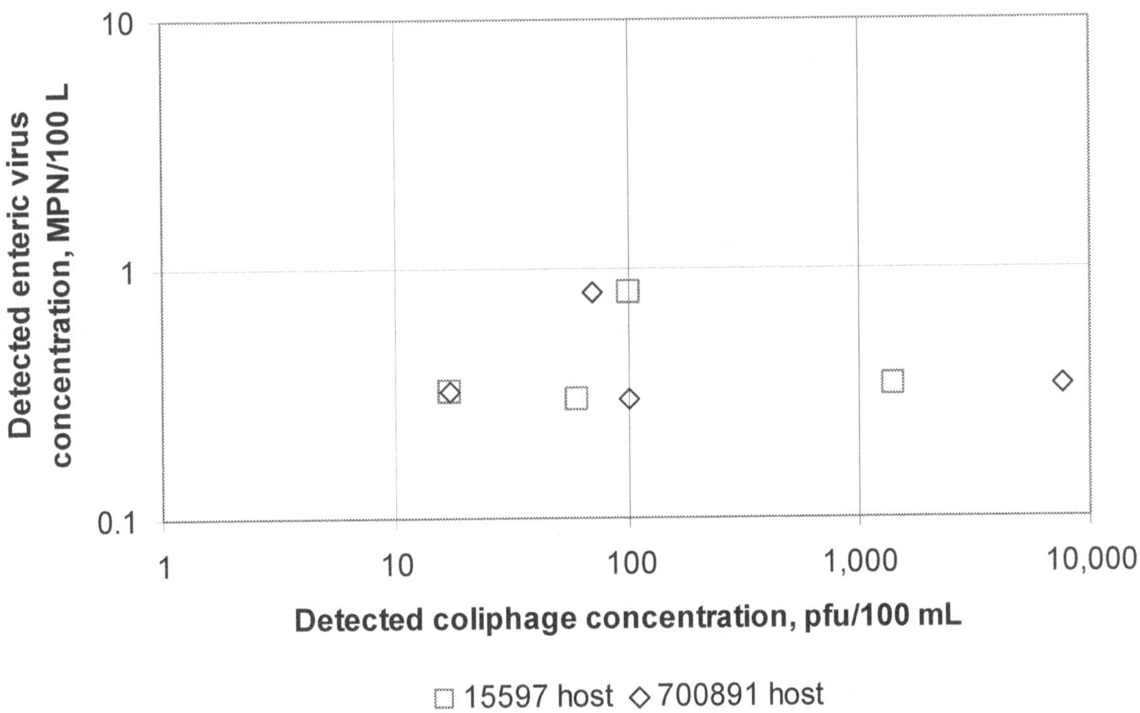

Figure 4-5. Scatter Plot of Detected Coliphage Concentrations (PFU•100 mL-1) vs. Detected Enterovirus Concentrations (MPN·100 L-1) in the Final Effluent from Six Wastewater Treatment Plants (n=8).. (◊ – Coliphage with E. coli ATCC 700891 as Host and □ – Coliphage with E. coli ATCC 15597 as Host).

4.2 Reduction of Indicators and Pathogens Through Wastewater Treatment

The persistence of bacteria, viruses, and protozoa is related to the initial concentration of each microbial species and the cumulative effectiveness of each sequential treatment step. Each successive treatment step (biological treatment, filtration, and disinfection) has the potential to reduce the concentrations of indicators and pathogens, depending on the wastewater characteristics, flowrate, operating conditions, and the overall treatment performance. If the treatment system is highly effective, the concentrations of indicators and pathogens may be below detection limits.

A comparison of the $\log_{10}$ reduction of indicators and pathogens from all facilities is shown in Figure 4-6. Data from all facilities were pooled and nondectected concentrations were reported as detection limits. As shown, for all cases except *Cryptosporidium*, biological treatment was effective for about 2 $\log_{10}$ reduction. Filtration was effective for about 0.3 to 1 $\log_{10}$ reduction for most of the measured parameters with higher removals associated with total and fecal coliforms and enterococci. Similar trends were observed for disinfection, however, the impacts of disinfection on protozoan pathogens are harder to elucidate since the monitored levels of *Giardia* and *Cryptosporidium* include total cysts or oocysts and do not reflect viability.

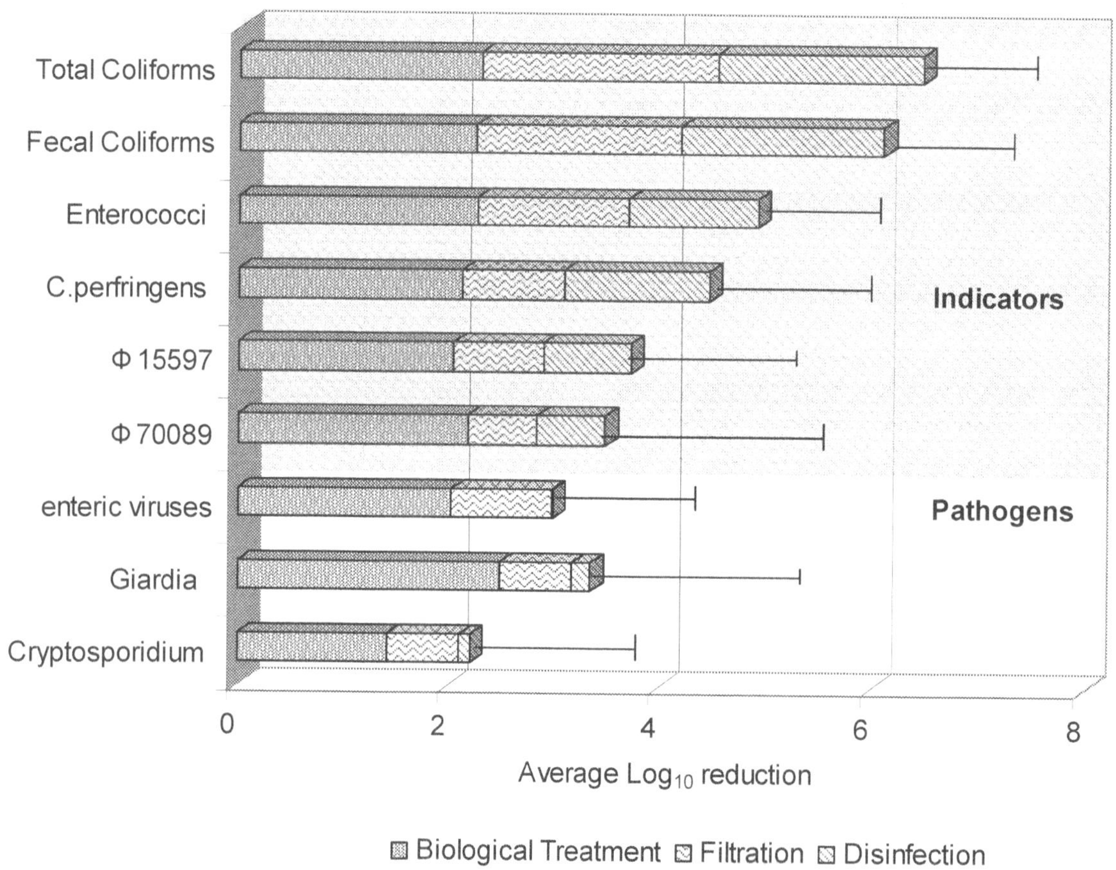

Figure 4-6. Mean $\log_{10}$ Reduction of Bacterial Indicators (cfu 100 mL-1), Coliphages (PFU 100 mL-1), Enteric Viruses (MPN 100 L-1), and Protozoa (Cysts or Oocysts 100 L-1) through Biological Treatment, Filtration, and Disinfection for Six Wastewater Reclamation Facilities. Error Bars Reflect Standard Deviation for each Indicator Through all Treatment Stages.

These results indicate that the proportion of samples containing pathogens is not consistent with the proportion of samples containing indicator organisms and thus indicator organism persistence was not an accurate predictor of the presence of pathogens in the final effluent samples across all six wastewater treatment plants.

CHAPTER 5.0

RECOMMENDATIONS AND CONCLUSIONS

5.1 Water Quality

- The concentrations of the traditional bacterial indicators (total and fecal coliforms), alternative indicators (Enterococci, coliphage) and pathogens were fairly consistent in all influents of the six treatment facilities and were not significantly different among the facilities (and were found at the most frequent and highest concentrations compared to post-treatment). Thus levels entering each facility did not ultimately influence the final water quality, which was influenced by processes, upstream to downstream, operations and design.
- Many wastewater treatment facilities are permitted to produce effluent that contains <1 **fecal coliform** bacteria/ 100 ml. Facilities A, B, C and F met this level in100% of disinfected effluent samples, while facilities D and E did so for only 66% and 33% of the samples. Only Facility C would have met the level of <1 **total coliform** bacteria /100ml in 75% of the samples collected. The other facilities had higher levels in 40-80% of the samples collected.
- Cultivatable enteric viruses and protozoan pathogens (*Giardia, Cryptosporidium*, total cysts and oocysts*)* were detected in untreated wastewater from all six facilities. Cultivatable enteric viruses were detected in 31% of the final effluent samples and were not detected in any of the samples from facilities B and E. Protozoan pathogens were detected in some of the effluent samples from all six facilities: *Giardia* cysts were detected in 79% of the final disinfected effluent samples while *Cryptosporidium* oocysts were detected in 39% of the final reclaimed effluents. Infectious *Cryptosporidium* oocysts were detected in 30% of the final reclaimed effluents and in four of the six facilities.
- About 10-30% of the *Cryptosporidium* oocysts remained infectious as determined via cell culture through biological treatment, filtration, and disinfection. *Giardia* cysts also remained viable (1-3% of the total IFA count) according to PI exclusion and DAPI inclusion. Infectious oocysts were not detected in any of the UV disinfected effluent samples,(detection limits of 2/100 L).
- The production of reclaimed water using secondary activated sludge processes, filtration and disinfection is not universally effective for removal of pathogens. Viruses ranged in concentration from 0.3 to 3.3 MPN PFU/100L, *Giardia* cysts ranged from 6 to 390/100 L and *Cryptosporidium* ranged from 4.6 to 114 oocysts/100 L in reclaimed water.

- Inefficiencies in secondary treatment and sedimentation result in a higher microbiological loading on filtration and may impact the effectiveness of filtration and disinfection for reduction bacteria, viruses, and protozoa.
- If coliform bacterial concentrations are used to assess microbiological water quality, facilities B and C outperformed the other facilities, while facilities A and F had intermediate performance, with D and E ranked as having the highest concentrations of these indicators. However, if detection of coliform bacteria is coupled with monitoring of enterococci, *Clostridium* and coliphage, Facilities F and E outperformed the other facilities, while A and B ranked in the middle and Facilities C and D had the highest frequency of detection of indicators at the detection limits used in this study.
- Based on the viruses, Facilities B, C, and E ranked as having the best water quality, D and F ranked in the middle and Facility A had the poorest quality. Based on *Giardia* cysts, Facility F had the best water quality, Facility F along with E and B also ranked the best based on *Cryptosporidium* oocysts. Facilities B, D, and E compared to A and D ranked in the middle for cysts and oocysts and Facilities A and C had the poorest water quality for cysts and Facility C based on oocysts.
- The use of traditional indicators as monitoring tools would not have identified the potential presence of pathogens in Facilities A and C.

5.2 Operations and Design

The data obtained in this study represented only six samples from each facility at each site after the treatment processes. Thus while this study represents the most comprehensive one of its type carried out to date, much more data are necessary to confirm the trends and results that have been observed here. The effluent quality for bacteria, coliphage, viruses and *Giardia* cysts from the different facilities was shown to be statistically significantly different, as the concentrations in the influent were not statistically different, this difference can be related to removal by the processes.

- Operation of biological treatment with higher levels of MLSS and longer MCRTs tended to result in increased removal of microbial indicators and pathogens.
- Prechlorinated shallow sand (effective size: 0.6 mm) filters were more effective than deep bed dual media (anthracite and >1 mm sand) or monomedia (anthracite or sand) filters for removal of bacterial indicators and viruses.
- Hydraulic loading and filter depths impacted removal of some of the indicators, with improved removal associated with lower hydraulic loading rates (0.6-2gpm/ft^2) and deeper filters (1-1.8m).
- UV inactivation (Facility E) of bacteria and coliphage indicators removed 92 to99% of these organisms but was not as effective as the extended contact times with chlorine which achieved 99.9-99.99% inactivation (Facilities C and B).
- Overall, Facility C with shortest MCRTs, anthracite monomedia filtration, and lowest range of chlorine residuals had the poorest finished water quality with respect to *Cryptosporidium*, coliphage, enterococci and *Clostridium*. This was despite the fact that the facility had the longest chlorine contact times.

- Facility A had the poorest finished water quality in regard to viruses and *Giardia* (poor removal of cysts). This facility had a cloth filter, highest loading rate and lowest range of chlorine contact times.
- Facilities D and E had the worst finished water quality in regard to total and fecal coliform bacteria. The UV disinfection at Facility E was ineffective at achieving the coliform standards of nondetects for the bacterial indicators although it appeared to be effective against the viruses and should have been effective based on design against the parasites. Facility D had poorer removal by the filtration but similar disinfection contact time as Facility A (which was the lowest of all facilities) thus producing a final water quality with greater concentrations of bacteria.
- Facility F produced the best overall microbial quality with the longest retention times, deepest filters (dual media), least amount of ammonia impacting disinfection (70-90 min. contact times with 4-6mg/L of residual chlorine) for parasites, viruses, and indicators. Despite this, viruses and infectious *Cryptosporidium* were still detected.
- The longer chlorine contact times in Facility B (due to prechlorination of the filter) and Facility C likely contributed to reductions of the viruses in those facilities and may be needed to achieve nondetectable levels of cultivatable viruses in the reclaimed water effluent (using current methods for virus detection).
- A high percentage of reclaimed water samples were positive for IFA total counts of cysts and oocysts, and some were positive for infectious *Cryptosporidium.* A rigorous risk assessment should be conducted to determine whether other barriers to reduce viable protozoa are needed.

Reclaimed water facilities can be designed and operated to produce a high water quality. Greater attention needs to be focused on improved secondary treatment in new facilities with longer retention and greater solids control. Filtration should be designed with greater media depths, however operationally efficiency can be improved with coagulation and prechlorination, which will enhance filtration/retention and inactivation, respectively. Finally, disinfection is effected by the upstream processes, chlorination and UV combinations provide barriers against viruses, bacteria and protozoa. Extended contact times improve reductions of viruses and bacteria and UV disinfection should be effective against the protozoa. Reclaimed facilities may take the secondary effluent from existing facilities without regard to microbial water quality. This study has demonstrated that the practice should be to monitor and examine indicators, alternative indicators, viruses and parasite in the effluent first. Then while it may be impossible to change the secondary treatment, greater attention can be paid to the designs and operations used for filtration and disinfection and their optimization given the quality of the effluent from the sewage treatment plant. Predisinfection to enhance virus and bacteria inactivation may be needed at reclaimed facilities that have little opportunity to change other design parameters.

Based on this study, it is evident that increased monitoring for alternative indicators and pathogens is warranted to build a better data base on the occurrence and reductions of enteric bacteria, viruses and parasites in reclaimed water. Finally the authors of this report believe that more discussion should be undertaken regarding the efficacy of process requirements versus water quality goals for reclaimed water at the national level. Reclaimed water as monitored in this study in Arizona, California, and Florida is not pathogen free and exposure of the public to these waters carries some risk, albeit this level may be very low and quite acceptable to most populations. Integration of microbiological monitoring with control factors associated with process design and operations may lead to a more robust approach for assuring the safety of reclaimed water.

CHAPTER 6.0

RECOMMENDATIONS FOR FURTHER RESEARCH

- More research is needed in detailing the removal of bacteria, parasites, and viruses by wastewater and reclaimed water facilities at the full-scale to examine the uncertainty and variability associated with the monitoring methods, quality of influents, treatment processes, seasonality, designs, and operations.
- While in theory UV disinfection was designed to inactivate the bacteria, Facility E with UV disinfection performed the poorest for coliform bacteria reductions. More work needs to be done on full-scale monitoring of UV systems. Adenoviruses are the most resistant to UV disinfection and are found in high concentrations in sewage, therefore wastewater treatment with UV should include an assessment of Adenoviruses in the future.
- For most waterborne pathogenic microorganisms, historical databases on occurrence in water do not exist or there are few. For example, all of the microorganisms on the U.S. EPA's "Contaminant Candidate List" require a database for an assessment on exposure to be adequate. Of the 13 possible organisms on the CCL, nine are associated with sewage and monitoring should be undertaken.
- More work should be focused on hydraulic retention and filter loading rates.
- Genetic characterization of *Cryptosporidium* as well as the bacteria should be undertaken for identification of key markers that could be used for source tracking in the future.
- There is a need for a practical measure of *Giardia* cyst viability, as the cysts are there in higher concentrations than often the viruses or oocysts, the ability to determine inactivation by disinfection is needed.
- The changes in water quality after treatment via storage and distribution and the need for monitoring water quality at the site of application should be evaluated.
- While some facilities/processes did not perform as well as others (shallow bed filters), there should be an assessment of whether these systems can change operations or need to be reconfigured.

APPENDIX OF DATA SUMMARIES

Table 1. Summary of concentrations of bacterial indicators, coliphages, enteric viruses, *Giardia*, and *Cryptosporidium* in untreated wastewater from treatment facilities A-F.

	INFLOW	Total Coliforms (cfu/100mL)	Fecal Coliforms (cfu/100mL)	Enterococci (cfu/100mL)	C.perfringens (cfu/100mL)	C. perfringens spores (cfu/100mL)	Coliphage - 15597 host (pfu/100mL)	Coliphage -F-amp host (pfu/100mL)	Enterovirus MPN/100L	Giardia (cysts/100L)	Cryptosporidium (oocysts/100L)	Infectious oocysts, #/100 L
plant a	n	5	5	5	5	5	5	5	5	5	5	5
	%pos	100	100	100	100	100	100	100	100	100	100	20
	average	3.64E+07	4.32E+06	3.45E+06	9.51E+03	1.01E+04	1.46E+05	1.58E+05	8.00E+02	3.37E+05	8.63E+03	4.83E+02
	geometric mean	3.49E+07	3.97E+06	2.41E+06	7.81E+03	7.74E+03	5.95E+04	6.45E+04	6.45E+02	2.62E+04	1.43E+03	2.33E+02
plant b	n	6	6	6	5		5	5	5	6	6	5
	%pos	100	100	100	100		100	100	60	100	83	20
	average	3.67E+07	3.56E+06	1.18E+06	1.70E+05		2.58E+05	3.08E+05	1.43E+04	8.60E+04	9.33E+03	1.95E+03
	geometric mean	3.01E+07	3.00E+06	7.05E+05	1.00E+05		1.97E+05	3.00E+05	5.41E+03	3.72E+04	3.78E+03	1.31E+03
plant c	n	5	5	6	5		5	5	5	5	5	5
	%pos	100	100	100	100		100	100	100	100	80	0
	average	4.52E+07	5.85E+06	2.35E+06	1.05E+06		2.26E+06	2.02E+07	1.86E+04	2.02E+05	3.67E+03	1.37E+03
	geometric mean	4.36E+07	5.31E+06	2.08E+06	9.68E+05		7.29E+05	8.79E+05	9.37E+03	7.94E+04	1.60E+03	7.08E+02
plant d	n	6	6	6	6	5	6	6	6	6	6	6
	%pos	100	100	100	67	80	100	100	67	100	50	33
	average	4.36E+07	3.40E+06	5.57E+05	1.50E+04	1.59E+04	2.46E+05	4.83E+04	2.06E+03	6.53E+04	7.52E+03	4.88E+02
	geometric mean	4.21E+07	3.27E+06	4.37E+05	2.00E+03	1.97E+04	1.25E+05	2.82E+04	1.21E+03	3.10E+04	1.33E+03	4.10E+02
plant e	n	5	5	5	4		5	5	5	5	5	4
	%pos	100	100	100	75		100	100	100	100	80	50
	average	3.00E+07	9.10E+06	1.59E+06	4.65E+04		9.35E+05	1.48E+06	7.94E+02	5.71E+04	5.84E+03	8.57E+02
	geometric mean	1.96E+07	3.56E+06	7.45E+05	5.73E+03		1.17E+05	6.07E+04	4.64E+02	1.95E+04	1.56E+03	8.05E+02
plant f	n	5	5	5	5		5	5	5	6	6	4
	%pos	100	100	100	100		100	100	100	100	67	25
	average	3.41E+07	2.64E+06	7.36E+05	1.40E+05		2.84E+05	3.14E+05	1.52E+04	3.05E+04	1.49E+03	4.37E+02
	geometric mean	3.32E+07	2.09E+06	5.98E+05	8.97E+04		1.86E+05	1.85E+05	7.16E+03	6.12E+03	6.55E+02	2.40E+02

Table 2. Summary of concentrations of bacterial indicators, coliphages, enteric viruses, *Giardia*, and *Cryptosporidium* in secondary effluents from treatment facilities A-F.

	Secondary Effluent	Total Coliforms (cfu/100mL)	Fecal Coliforms (cfu/100mL)	Enterococci (cfu/100mL)	C.perfringens (cfu/100mL)	C. perfringens spores (cfu/100mL)	Coliphage - 15597 host (pfu/100mL)	Coliphage -F- amp host (pfu/100mL)	Enterovirus MPN/100L	Giardia (cysts/100L)	Cryptosporidium (oocysts/100L)	Infectious oocysts, #/100 L
plant a	n	5	5	5	5	5	5	5	5	5	5	5
	%pos	100	100	100	60	60	80	80	40	80	100	20
	average	7.05E+05	7.39E+04	1.64E+04	3.29E+02	3.13E+02	2.06E+03	3.74E+02	1.14E+01	3.69E+03	1.50E+02	1.03E+01
	geometric mean	4.67E+05	4.79E+04	1.15E+04	6.37E+01	6.54E+01	4.36E+02	1.76E+02	5.85E+00	7.56E+02	8.39E+01	1.00E+01
plant b	n	5	5	6	5		5	5	5	6	6	4
	%pos	100	100	100	100		100	100	80	100	100	25
	average	4.13E+05	6.22E+04	9.62E+03	3.06E+03		2.33E+03	7.20E+02	4.12E+01	1.67E+02	6.76E+01	4.64E+01
	geometric mean	3.91E+05	5.47E+04	7.93E+03	2.89E+03		2.07E+03	5.08E+02	2.72E+01	8.34E+01	4.52E+01	
plant c	n	5	5	6	5		5	5	5	6	6	4
	%pos	100	100	100	100		100	100	100	100	83	25
	average	5.85E+05	2.27E+05	7.80E+04	1.05E+04		2.00E+07	7.90E+03	1.66E+02	1.77E+03	2.25E+02	2.48E+01
	geometric mean	5.04E+05	1.28E+05	5.28E+04	5.74E+03		1.49E+05	4.70E+03	1.33E+02	1.68E+02	4.87E+01	2.12E+01
plant d	n	7	7	7	7	5	6	6	6	6	6	6
	%pos	86	86	71	71	80	100	83	67	100	83	0
	average	5.74E+04	1.65E+04	2.94E+03	1.03E+03	1.94E+03	2.62E+03	7.45E+02	5.02E+00	1.45E+03	7.26E+01	9.89E+00
	geometric mean	1.21E+04	2.53E+03	6.32E+02	6.29E+01	1.50E+02	5.64E+02	3.69E+02	4.35E+00	9.74E+01	2.79E+01	9.75E+00
plant e	n	5	5	5	4		5	5	5	5	5	4
	%pos	100	100	100	50		100	80	60	80	40	25
	average	1.33E+06	3.01E+04	7.49E+03	1.84E+02		1.23E+04	8.19E+02	7.18E+00	9.28E+01	3.72E+01	1.20E+01
	geometric mean	1.85E+05	1.67E+04	3.89E+03	7.95E+01		1.01E+03	2.66E+02	2.99E+00	6.55E+01	3.09E+01	1.13E+01
plant f	n	5	5	5	5		5	6	6	5	5	5
	%pos	100	100	100	100		80	83	67	100	80	20
	average	1.67E+05	4.07E+03	8.99E+02	2.28E+02		7.10E+01	7.17E+01	5.66E+02	5.59E+02	5.88E+01	4.46E+01
	geometric mean	1.28E+05	3.23E+03	6.99E+02	1.68E+02		4.64E+01	3.80E+01	4.42E+01	3.60E+02	5.10E+01	3.56E+01

Table 3. Summary of concentrations of bacterial indicators, coliphages, enteric viruses, *Giardia,* and *Cryptosporidium* in filtered secondary effluents from treatment facilities A-F.

	Filtered effluent	Total Coliforms (cfu/100mL)	Fecal Coliforms (cfu/100mL)	Enterococci (cfu/100mL)	C.perfringens (cfu/100mL)	C. perfringens spores (cfu/100mL)	Coliphage - 15597 host (pfu/100mL)	Coliphage -F-amp host (pfu/100mL)	Enterovirus MPN/100L	Giardia (cysts/100L)	Cryptosporidium (oocysts/100L)	Infectious oocysts, #/100 L
plant a	n	5	5	5	5	5	5	5	5	5	5	5
	%pos	80	60	80	40	40	40	40	60	100	100	40
	average	5.32E+02	4.13E+02	2.08E+02	5.88E+01	3.60E+01	4.74E+02	1.84E+02	4.31E+00	4.05E+02	7.66E+01	3.58E+00
	geometric mean	7.54E+01	3.68E+01	4.02E+01	4.22E+00	3.74E+00	6.59E+01	4.50E+01	1.87E+00	2.45E+02	3.46E+01	3.38E+00
plant b	n	6	6	6	5		6	6	5	5	5	4
	%pos	83	17	83	100		100	100	20	100	80	25
	average	1.06E+03	9.25E+02	2.10E+02	3.28E+01		2.10E+02	2.34E+02	1.98E+00	3.87E+01	6.92E+00	4.33E+00
	geometric mean	1.65E+01	1.59E+00	6.81E+00	2.60E+01		1.04E+02	1.04E+02	2.38E+00	2.40E+01	5.71E+00	3.67E+00
plant c	n	6	6	6	5		5	5	5	6	6	4
	%pos	100	67	67	60		100	100	80	83	67	25
	average	2.21E+04	7.31E+02	1.79E+04	7.36E+03		1.55E+04	3.87E+03	2.63E+01	1.99E+01	2.06E+02	3.48E+00
	geometric mean	6.85E+03	4.37E+02	1.89E+03	6.88E+02		7.31E+03	2.02E+03	1.93E+01	8.35E+00	1.77E+01	3.38E+00
plant d	n	7	7	7	7	5	7	7	6	6	6	6
	%pos	100	100	100	71	80	14	14	67	50	67	17
	average	1.54E+04	9.54E+03	2.56E+03	3.94E+01	3.65E+01	1.09E+02	1.68E+02	2.67E+00	4.35E+01	6.27E+01	4.37E+00
	geometric mean	6.85E+03	2.21E+03	1.22E+03	1.27E+01	1.82E+01	1.87E+01	2.00E+01	2.34E+00	1.45E+01	2.15E+01	4.23E+00
plant e	n	5	5	5	4		5	5	5	5	5	4
	%pos	100	100	100	75		100	100	40	100	40	25
	average	1.79E+04	2.02E+04	5.80E+02	8.18E+01		1.14E+03	1.41E+02	6.98E-01	3.64E+01	1.37E+01	2.40E+00
	geometric mean	4.83E+03	2.20E+03	4.18E+02	2.64E+01		3.33E+02	1.34E+02	5.75E-01	1.45E+01	6.57E+00	1.74E+00
plant f	n	5	6	5	5		6	5	6	6	6	4
	%pos	100	67	80	100		67	40	67	100	67	0
	average	1.55E+02	7.81E+01	8.41E+01	3.50E+01		3.10E+01	2.50E+01	5.48E+00	1.79E+01	7.15E+00	2.83E+00
	geometric mean	1.09E+02	1.27E+01	3.14E+01	2.46E+01		2.11E+01	1.60E+01	2.80E+00	1.02E+01	4.70E+00	2.68E+00

Table 4. Summary of concentrations of bacterial indicators, coliphages, enteric viruses, *Giardia*, and *Cryptosporidium* in disinfected, filtered secondary effluents from treatment facilities A-F.

	Disinfected effluent	Total Coliforms (cfu/100mL)	Fecal Coliforms (cfu/100mL)	Enterococci (cfu/100mL)	C.perfringens (cfu/100mL)	C. perfringens spores (cfu/100mL)	Coliphage - 15597 host (pfu/100mL)	Coliphage -F- amp host (pfu/100mL)	Enterovirus MPN/100L	Giardia (cysts/100L)	Cryptosporidium (oocysts/100L)	Infectious oocysts, #/100 L
plant a	n	5	5	5	5	5	5	5	5	5	5	5
	%pos	60	40	0	60	40	20	60	60	100	100	20
	average	3.47E+00	2.53E-01	2.00E-01	5.35E+00	4.48E+00	2.00E+01	3.40E+01	3.31E+00	3.89E+02	1.27E+01	2.54E+00
	geometric mean	1.01E+00	2.45E-01	2.00E-01	1.35E+00	1.07E+00	1.43E+01	2.27E+01	9.11E-01	1.87E+02	9.77E+00	1.87E+00
plant b	n	6	6	6	5		5	5	5	6	6	4
	%pos	50	0	33	80		100	100	0	100	83	25
	average	1.17E+00	2.83E-01	3.73E+00	7.04E+00		2.60E+01	6.20E+01	8.30E-01	2.31E+01	4.58E+00	1.04E+01
	geometric mean	8.09E-01	2.56E-01	7.79E-01	4.16E+00		2.17E+01	3.87E+01	7.23E-01	1.34E+01	3.76E+00	5.84E+00
plant c	n	4	4	4	4		4	4	4	4	4	2
	%pos	50	0	75	100		75	75	25	75	75	0
	average	2.68E+00	3.00E-01	1.54E+01	1.68E+01		3.65E+02	2.08E+03	5.85E-01	2.77E+01	1.14E+02	2.20E+00
	geometric mean	7.85E-01	2.63E-01	5.47E+00	8.25E+00		4.86E+01	1.72E+02	4.55E-01	1.06E+01	3.35E+01	2.20E+00
plant d	n	6	6	6	6	5	6	6	6	6	6	6
	%pos	83	50	50	50	60	0	17	17	83	83	33
	average	3.35E+02	3.50E+02	3.11E+02	1.73E+00	1.15E+00	1.00E+01	1.17E+01	3.12E-01	2.23E+01	4.92E+01	6.64E+00
	geometric mean	7.95E+00	3.73E+00	5.32E+00	5.41E-01	9.44E-01	1.00E+01	1.12E+01	3.10E-01	1.39E+01	2.12E+01	5.19E+00
plant e	n	5	5	5	4		5	5	5	5	5	4
	%pos	80	80	20	25		0	0	0	80	20	0
	average	1.19E+01	1.87E+02	4.38E+00	4.00E-01		1.00E+01	1.00E+01	5.22E-01	1.97E+01	6.50E+00	1.74E+00
	geometric mean	5.31E+00	2.32E+00	5.50E-01	2.99E-01		1.00E+01	1.00E+01	4.62E-01	1.11E+01	4.94E+00	1.69E+00
plant f	n	5	5	5	5		5	5	5	6	6	4
	%pos	60	0	0	40		80	80	80	50	67	50
	average	1.13E+01	2.00E-01	2.00E-01	1.56E+00		1.04E+01	1.04E+01	3.36E-01	6.70E+00	6.50E+00	8.33E+00
	geometric mean	1.17E+00	2.00E-01	2.00E-01	6.25E-01		9.68E+00	9.68E+00	3.33E-01	4.53E+00	3.53E+00	4.14E+00

References

Adams, M.H. 1959. *Bacteriophages*, Interscience Publishers, New York

American Public Health Association, American Water Works Association and Water Environment Federation. 1998. Standard Methods for the Examination of Water and Wastewater, 20th Edition. New York.

Arizona. 1984. Water Quality Standards of Wastewater Reuse. Arizona, Department of State, Phoenix. AZ

Bisson, J.W. and V.J. Cabelli. 1979. Membrane filter enumeration method for *Clostridium perfringens*. Appl.Environ. Microbiol. 37:55-66.

Black, E.K., G.R. Finch, R. Taghi-Kilami, M. Belosevic.1996. Comparison of assays for Cryptosporidium parvum oocysts viability after chemical disinfection. FEMS Microbiol. Lett. 135, 187-189.

Campbell, A.T., L.J. Robertson, and H.V. Smith. 1992. Viability of *Cryptospordium parvum* oocysts—correlation of in vitro excystation with inclusion or exclusion of fluorogenic vital dyes. Appl. Environ. Microbiol. 58:3488-3493.

County Sanitation Districts of Los Angeles County (CSDOLAC) 1977. *Pomona Virus Study, Final Report*, Prepared for California State Water Resources Control Board and U.S Environmental Protection Agency.

Crook, J. 2003. *Summary of Water Reuse Criteria*. CH2M HILL Technical Memorandum prepared for the Massachusetts Department of Environmental Protection, May 5, 2003.

Gardner, T.B. and D.R. Hill. 2001. Treatment of giardiasis. Clin Microbiol Rev. 14:114-28

Haas, C.H., J.B. Rose, and C.P. Gerba (eds). 1999. *Quantitiative Microbial Risk Assessment*. John Wiley and Sons, New York, NY.

Harwood, V.J. 2001. Detection and occurrence of indicator organisms and pathogens: Literature Review. Water Environment Research 73 (5):

Havelaar, A.H. 1993. F-specific RNA bacteriophages are adequate model organisms for enteric viruses in fresh water. Appl. Environ. Microbiol. 59:2956-2962

Jenkins, M.B., L.J. Anguish, D.D. Bowman, M.J. Walker, and W.C. Ghiorse. 1997. Assessment of a dye permeability assay for determination of inactivation rates of *Cryptosporidium parvum* oocysts. Appl. Environ. Microbiol. 63:3844-3850.

Leclerc, H., D.A. Mossel, S.C. Edberg, and C.B. Struijk**.** 2001. Advances in the bacteriology of the coliform group: their suitability as markers of microbial water safety. Annual Review of Microbiology 55:201-34.

Levine, A.D. and T. Asano. 2004 Recovering sustainable water from wastewater, Environ. Sci. Technol.38:201A-208A

Mac Kenzie, W.R., N.J. Hoxie, M.E. Proctor, M.S. Gradus, K.A. Blair, D.E. Peterson, J.J. Kazmierczak, D.G. Addiss, K.R. Fox, J.B. Rose, and J.P. David. 1994. A massive outbreak in Milwaukee of *Cryptosporidium* infection transmitted through the public water supply. *New Engl. J. Med.* 331(3): 161-167.

Mara, D. and N. Horan.(eds.). 2003. *Handbook of Water and Wastewater Microbiology*. New York, NY. Pg. 193-208.

Murray, P., K.S. Rosenthal, G.S. Kobayashi, M.A. Pfaller. 2001. Medical Microbiology. Mosby-Year Book, Inc., St. Louis, MO.

Neumann, N.F., L.L. Gyurek, L. Gammie, G.R. Finch, and M. Belosevic. 2000. Comparison of animal infectivity and nucleic acid staining for assessment of *Cryptosporidium parvum* viability in water. Appl. and Environ. Microbiol. 66:406-412.

Rose, J.B. 1986. Microbial aspects of wastewater reuse for irrigation. *CRC Critical Reviews in Environmental Control.* 16: 231-256.

Rose J.B., L.J. Dickson, S.R. Farrah, and R.P. Carnahan. 1996. Removal of pathogenic and indicator microorganisms by a full-scale water reclamation facility. *Wat. Res.* 30(11): 2785-2797.

Rose, J.B., S.R. Farrah, D.E. Friedman, K. Riley, C.L. Hamann, and M. Robbins. 1999. Public health evaluation of advanced reclaimed water for potable applications. *Wat. Sci. Tech.* 40(4-5):247-252.

Rose, J.B., D.E. Huffman, K. Riley, S.R. Farrah, J.O. Lukasik, and C.L. Harman. 2001 Reduction of enteric microorganisms at the Upper Occoquan Sewage Authority water reclamation plant. *Wat. Environ. Res.* 73:711-720

Rose, J.B. D.E. Huffman, and A. Gennaccaro, 2002. Risk and control of waterborne cryptosporidiosis. *FEMS Microbiology Reviews* 26:113-123.

Sauch, J.F., D. Flanigan, M.L. Galvin, D. Berman, and W. Jakubowski. 1991. Propidium iodide as an indicator of *Giardia* cyst viability. Appl. Environ. Microbiol. 57:3243-3247.

Schupp, D.G. and S.L. Erlandsen. 1987. A new method to determine *Giardia* cyst viability - correlation of fluorescein diacetate and propidium iodide staining with animal infectivity. Appl. Environ. Microbiol. 53:704-707.

Slifko T.R., D. Friedman, J.B. Rose, and W. Jankubowski. 1997. An in vitro method for detecting infectious *Crytosporidium* oocysts with cell culture. *Appl. Environ. Microbiol.* 63:3669-3675.

Slifko, T.R., D.E. Huffman, and J.B. Rose. 1999. A most probable assay for enumeration of infectious *Cryptosporidium parvum* oocycts. *Appl. Environ. Microbiol*. 65:3936-3941.

Smith, H.V. and J.B. Rose. 1998. Waterborne cryptosporidiosis current status. *Parasitology Today.* 14:14-22.

Smith, H.V. 1996. Detection of Cryptosporidium and Giardia in Water. In Pickup, R.W., Saunders, J.R. (Eds.), Molecular Approaches to Environmental Microbiology. Ellis-Horwood Ltd., Hemel Hempstead, UK. pp.195-225

Sobsey, M.D. 1989. Inactivation of health-related microorganisms in water by disinfection processes. *Water Sci. Technol.* 21:179-195.

State of California 1978. Administrative Code, Title 22, Division 4, Environmental Health, Chapter 3 *Wastewater Reclamation Criteria*, Department of Health Services, pp. 1557-1610.

Turner, S.J. and G.D. Lewis. 1995. Comparison of F-specific bacteriophage, enterococci, and faecal coliform densitiesthrough a wastewater treatment process employing oxidation ponds. Water Science and Technology 31:85-89.

U.S. Environmental Protection Agency. 1992. Manual guidelines for water reuse. U.S. EPA/625/R-92/004.

U.S. Environmental Protection Agency. 1996. ICR Microbial Laboratory Manual EPA/600/R-95/178. U.S. Environmental Protection Agency.

U.S. Environmental Protection Agency. 1995. ICR Protozoan Method for Detecting Giardia Cysts and Cryptosporidium Oocysts in Water by a Fluorescent Antibody Procedure. U.S. EPA/814-B-95-003.

U.S. Environmental Protection Agency. 1997. Method 1600: membrane filter test methods for enterococci in water. EPA-821/R-97/004. U.S. Environmental Protection Agency.

U.S. Environmental Protection Agency. 1999. Method 1623: *Cryptosporidium* and *Giardia* in Water by Filtration/IMS/FA. Office of Water. EPA/821/R-99/006.

Yanko, W.A., J.L. Jackson, F.P. Williams, A.S. Walker, M. Castillo. 1999. An unexpected temporal pattern of coliphage isolation in groundwaters sampled from wells at varied distances from reclaimed water recharge sites. *Wat. Res.* 33(1): 53-64.

York, D.W. and L. Walker-Coleman. 2000. Pathogen standards for reclaimed water, Water Environ. Technol.12:59-61.

WASTEWATER UTILITY

Alabama
Montgomery Water Works & Sanitary Sewer Board

Alaska
Anchorage Water & Wastewater Utility

Arizona
Gila Resources

Glendale, City of, Utilities Department

Mesa, City of

Peoria, City of

Pima County Wastewater Management

Arkansas
Little Rock Wastewater Utility

California
Calaveras County Water District

Central Contra Costa Sanitary District

Contra Costa Water District

Crestline Sanitation District

Delta Diablo Sanitation District

Dublin San Ramon Services District

East Bay Dischargers Authority

East Bay Municipal Utility District

El Dorado Irrigation District

Fairfield-Suisun Sewer District

Fresno Department of Public Utilities

Irvine Ranch Water District

Las Virgenes Municipal Water District

Livermore, City of

Lodi, City of

Los Angeles, City of

Los Angeles County, Sanitation Districts of

Napa Sanitation District

Orange County Sanitation District

Palo Alto, City of

Riverside, City of

Sacramento Regional County Sanitation District

San Diego Metropolitan Wastewater Department, City of

San Francisco, City & County of

San Jose, City of

Santa Barbara, City of

Santa Rosa, City of

South Bayside System Authority

South Coast Water District

South Orange County Wastewater Authority

Sunnyvale, City of

Union Sanitary District

West Valley Sanitation District

Colorado
Boulder, City of

Colorado Springs Utilities

Littleton/Englewood Water Pollution Control Plant

Metro Wastewater Reclamation District, Denver

Connecticut
The Mattabasset District

New Haven, City of, WPCA

District of Columbia
District of Columbia Water & Sewer Authority

Florida
Broward, County of

Fort Lauderdale, City of

Gainesville Regional Utilities

JEA

Miami-Dade Water & Sewer Authority

Orange County Utilities Department

Orlando, City of

Reedy Creek Improvement District

Seminole County Environmental Services

St. Petersburg, City of

Stuart Public Utilities

Tallahassee, City of

Tampa, City of

Toho Water Authority

West Palm Beach, City of

Georgia
Atlanta Department of Watershed Management

Augusta, City of

Clayton County Water Authority

Cobb County Water System

Columbus Water Works

Fulton County

Gwinnett County Department of Public Utilities

Savannah, City of

Hawaii
Honolulu, City and County of

Illinois
American Bottoms Wastewater Treatment Plant

Greater Peoria Sanitary District

Kankakee River Metropolitan Agency

Metropolitan Water Reclamation District of Greater Chicago

North Shore Sanitary District

Wheaton Sanitary District

Iowa
Ames, City of

Cedar Rapids Wastewater Facility

Des Moines Metro Wastewater Reclamation Authority

Iowa City, City of

Kansas
Johnson County Unified Wastewater Districts

Unified Government of Wyandotte County/Kansas City, City of

Kentucky
Louisville & Jefferson County Metropolitan Sewer District

Louisiana
Sewerage & Water Board of New Orleans

Maine
Bangor, City of

Portland Water District

Maryland
Anne Arundel County Bureau of Utility Operations

Howard County Department of Public Works

Washington Suburban Sanitary Commission

Massachusetts
Boston Water & Sewer Commission

Upper Blackstone Water Pollution Abatement District

Michigan
Ann Arbor, City of

Detroit, City of

Holland Board of Public Works

Lansing, City of

Owosso Mid-Shiawassee County WWTP

Saginaw, City of

Wayne County Department of Environment

Wyoming, City of

Minnesota
Rochester, City of

Western Lake Superior Sanitary District

Missouri
Independence, City of

Kansas City Missouri Water Services Department

Little Blue Valley Sewer District

Nebraska
Lincoln Wastewater System

Nevada
Henderson, City of

New Jersey
Bergen County Utilities Authority

Passaic Valley Sewerage Commissioners

New York
New York City Department of Environmental Protection

Rockland County Solid Waste Management Authority/ Sewer District

North Carolina
Charlotte/Mecklenburg Utilities

Durham, City of

Metropolitan Sewerage District of Buncombe County

Orange Water & Sewer Authority

Ohio
Akron, City of

Butler County Department of Environmental Services

Columbus, City of

Metropolitan Sewer District of Greater Cincinnati

Northeast Ohio Regional Sewer District

Summit, County of

Oklahoma
Tulsa, City of

Oregon
Clean Water Services

Eugene/Springfield Water Pollution Control

Water Environment Services

Pennsylvania
Philadelphia, City of

University Area Joint Authority, State College

South Carolina
Charleston Commissioners of Public Works

Mount Pleasant Waterworks & Sewer Commission

Spartanburg Sanitary Sewer District

Tennessee
Cleveland, City of

Knoxville Utilities Board

Murfreesboro Water & Sewer Department

Nashville Metro Water Services

Texas
Austin, City of

Dallas Water Utilities

Denton, City of

El Paso Water Utilities

Fort Worth, City of

Gulf Coast Waste Disposal Authority

Houston, City of

San Antonio Water System

Trinity River Authority

WERF SUBSCRIBERS

Utah
Salt Lake City Corporation

Virginia
Alexandria Sanitation Authority
Arlington, County of
Fairfax County Virginia
Hampton Roads Sanitation District
Henrico, County of
Hopewell Regional Wastewater Treatment Facility
Loudoun County Sanitation Authority
Lynchburg Regional WWTP
Prince William County Service Authority
Richmond, City of
Rivanna Water & Sewer Authority

Washington
Edmonds, City of
Everett, City of
King County Department of Natural Resources
Seattle Public Utilities
Sunnyside, Port of
Yakima, City of

Wisconsin
Green Bay Metro Sewerage District
Madison Metropolitan Sewerage District
Milwaukee Metropolitan Sewerage District
Racine, City of
Sheboygan Regional Wastewater Treatment
Wausau Water Works

Australia
South Australian Water Corp.
Sydney Water Corp.
Water Corp. of Western Australia

Canada
Greater Vancouver Regional District
Toronto, City of, Ontario
Winnipeg, City of, Manitoba

Mexico
Servicios de Agua y Drenaje de Monterrey, I.P.D.

New Zealand
Watercare Services Limited

Singapore
Singapore Public Utilities Board

United Kingdom
Yorkshire Water Services Limited

STORMWATER UTILITY

California
Los Angeles, City of, Department of Public Works
Monterey, City of
San Francisco, City & County of
Santa Rosa, City of
Sunnyvale, City of

Colorado
Boulder, City of

Georgia
Griffin, City of

Iowa
Cedar Rapids Wastewater Facility
Des Moines Metro Wastewater Reclamation Authority

Kansas
Overland Park, City of

Kentucky
Louisville & Jefferson County Metropolitan Sewer District

Maine
Portland Water District

Minnesota
Western Lake Superior Sanitary District

North Carolina
Charlotte, City of, Stormwater Services

Pennsylvania
Philadelphia, City of

Tennessee
Chattanooga Storm Water Management

Washington
Seattle Public Utilities

STATE

Arkansas Department of Environmental Quality
Fresno Metropolitan Flood Control District, Calif.
Urban Drainage & Flood Control District, Colo.

CORPORATE

ADS Environmental Services
The ADVENT Group Inc.
Alan Plummer & Associates
Alden Research Laboratory
Aqua-Aerobic Systems, Inc.
Aquateam–Norwegian Water Technology Centre A/S
BaySaver Inc.
BioVir Laboratories, Inc.
Black & Veatch
Boyle Engineering Corporation
Brown & Caldwell
Burns & McDonnell
CABE Associates Inc.
The Cadmus Group
Camp Dresser & McKee Inc.
Carollo Engineers Inc.
Carpenter Environmental Associates Inc.
CDS Technologies Inc.
Chemtrac Systems Inc.
CH2M HILL
Damon S. Williams Associates, LLC
David L. Sheridan, P.C.
Dewling Associates, Inc.
Earth Tech Inc.
Eco-Matrix
Ecovation
Environmental Engineers International
EMA Inc.
The Eshelman Company Inc.
Finkbeiner, Pettis, & Strout Inc.
Freese & Nichols, Inc.
ftn Associates Inc.
Fuss & O'Neill, Inc.
Gannett Fleming Inc.
Golder Associates Inc.
Greeley and Hansen LLC
Hazen & Sawyer, P.C.
HDR Engineering Inc.
HNTB Corporation
HydroQual Inc.
Infilco Degremont Inc.
Ingersoll-Rand Energy Systems
Insituform Technologies Inc.
Institute for Environmental Technology & Industry, Korea
Jacobson Helgoth Consultants Inc.
Jason Consultants Inc.
Jordan, Jones, & Goulding Inc.
KCI Technologies Inc.
Kelly & Weaver, P.C.
Kennedy/Jenks Consultants
Komline Sanderson Engineering Corporation
Lawler, Matusky & Skelly Engineers, LLP
Limno-Tech Inc.
Lombardo Associates Inc.
Malcolm Pirnie Inc.
McKim & Creed
MEC Analytical Systems Inc.
Metcalf & Eddy Inc.
MWH
New England Organics
Odor & Corrosion Technology Consultants Inc. (OCTC)
Oswald Green, LLC
PA Government Services Inc.
Parametrix Inc.
Parsons
Post, Buckley, Schuh & Jernigan
R&D Engineering/Conestoga Rover & Associates
The RETEC Group
RMC, Inc. (Raines, Melton & Carella)
R.M. Towill Corporation
Ross & Associates Ltd.
Royce Technologies
SAIC Maritime Technical Group
Short Elliott Hendrickson, Inc.
Stantec Consulting Group, Inc.
Stormwater Management, Inc.
Synagro Technologies, Inc.
Tetra Tech Inc.
Trojan Technologies Inc.
URS Corporation
Veolia Water NATC
Wade-Trim Inc.
Weston Solutions Inc.
Woodard & Curran
WRc/D&B, LLC
WWETCO, LLC
Zoeller Pump Company

INDUSTRY

American Electric Power
ChevronTexaco Energy Research & Technology Company
The Coca-Cola Company
Dow Chemical Company
DuPont Company
Eastman Kodak Company
Eli Lilly & Company
Merck & Company Inc.
ONDEO Services
Procter & Gamble Company
PSEG Services Corp.
RWE Thames Water Plc
Severn Trent Services Inc.
United Water Services LLC

Note: List as of 7/1/04

WERF Product Order Form

As a benefit of joining the Water Environment Research Foundation, subscribers are entitled to receive one complimentary copy of all final reports and other products. Additional copies are available at cost (usually $10). To order your complimentary copy of a report, please write "free" in the unit price column. WERF keeps track of all orders. If the charge differs from what is shown here, we will call to confirm the total before processing.

Name Title

Organization

Address

City State Zip Code Country

Phone Fax Email

Stock #	Product	Quantity	Unit Price	Total
			Postage & Handling	
			VA Residents Add 4.5% Sales Tax	
			Canadian Residents Add 7% GST	
			TOTAL	

Method of Payment:

(All orders must be prepaid.)

❑ Check or Money Order Enclosed

❑ Visa ❑ Mastercard ❑ American Express

Account No. Exp. Date

Signature

Shipping & Handling:

Amount of Order	United States	Canada & Mexico	All Others
Up to but not more than:	Add:	Add:	Add:
$20.00	$5.00*	$8.00	50% of amount
30.00	5.50	8.00	40% of amount
40.00	6.00	8.00	
50.00	6.50	14.00	
60.00	7.00	14.00	
80.00	8.00	14.00	
100.00	10.00	21.00	
150.00	12.50	28.00	
200.00	15.00	35.00	
More than $200.00	Add 20% of order	Add 20% of order	

*minimum amount for all orders

Note: Please make checks payable to the Water Environment Research Foundation.

To Order (Subscribers Only):

Log on to www.werf.org and click on the "Product Catalog."

Phone: (703) 684-2470
Fax: (703) 299-0742.

WERF
Attn: Subscriber Services
635 Slaters Lane
Alexandria, VA 22314-1177

To Order (Non-Subscribers):

Non-subscribers may be able to order WERF publications either through WEF (www.wef.org) or IWAP (www.iwapublishing.com). Visit WERF's website at www.werf.org for details.

IWA Publishing's authorised EU representative for General Product Safety Regulations is Diane D'Arras, 15 rue Duret, 75116 Paris, France, e-mail: safety@iwap.co.uk.

Printed and bound by CPI Group (UK) Ltd, Croydon, CR0 4YY
12/05/2026
02108879-0007